T0296125

Cambridge Tracts in Mathematics and Mathematical Physics

No. 14

THE TWISTED CUBIC

THE TWISTED CUBIC

WITH SOME ACCOUNT OF THE METRICAL PROPERTIES OF

THE CUBICAL HYPERBOLA

by

P. W. WOOD, M.A.
Fellow of Emmanuel College, Cambridge

Cambridge:
at the University Press

1913

CAMBRIDGE
UNIVERSITY PRESS

University Printing House, Cambridge CB2 8BS, United Kingdom

Cambridge University Press is part of the University of Cambridge.

It furthers the University's mission by disseminating knowledge in the pursuit of education, learning and research at the highest international levels of excellence.

www.cambridge.org
Information on this title: www.cambridge.org/9781107493728

First published 1913
Re-issued 2015

A catalogue record for this publication is available from the British Library

ISBN 978-1-107-49372-8 Paperback

PREFACE

THE plane curve of order two (the conic) has received its share of attention from analysts and geometers alike, but the obvious analogue, the space curve of order three (the twisted cubic), has been to some extent neglected by the analyst, and, so far as I know, there is no connected analytical treatment of its properties which assumes no previous acquaintance with the curve. It is hoped that the present tract will help to supply this defect.

Attention was first directed to the twisted cubic by Möbius in 1827, and ten years later Chasles in his *Aperçu Historique* urged the importance of a close study of the curve. Since then many writers have contributed memoirs, some synthetic in character, others taking the twisted cubic as a convenient medium for the interpretation of the invariant algebra of binary forms, comparatively few dealing with its metrical properties.

The first section, which treats of the projective properties of the curve, forms an introduction* to the second, which deals mainly with the relations between the asymptotes, the diameters and other elements associated with the cubical hyperbola. It is a matter of experience that the analytical discussion of descriptive properties is achieved most easily by the use of homogeneous coordinates, while metrical properties associate themselves naturally with Cartesian coordinates. The properties and formulae of the first section might have been re-discovered in terms of the Cartesian coordinates used in the second section, but no useful purpose would be served by this duplication of results. It should be remarked that, if imaginary elements are admitted, the properties of the cubical hyperbola do not differ essentially from those of the general cubical ellipse.

* The second section does not assume any knowledge of the results of §§ 15—25.

It will be seen that there are not many metrical properties involving reference to the imaginary circle at infinity; the cubical hyperbola does not in fact appear to lend itself to intimate association with the imaginary circle.

Analytical methods have been used throughout in preference to synthetical, and it will be understood that only the fringe of the subject can be touched upon. Some space has been economized by suppressing elementary algebraical reductions and by making no attempt to ascribe every result to its original author.

Nearly all the work has been done in collaboration with Mr G. T. Bennett and most of the novelties are due to him. A detailed account of my indebtedness would be tedious, but I would mention in particular the form of the conditions of § 15, the discovery of the mutually inscribed tetrahedra of § 22 with the allied results of §§ 23—25, the form of the treatment of the cubical hyperbola in §§ 28, 31, and the properties of the screw (p. 78) associated with any three lines.

It may be necessary to add that the remaining types of twisted cubic have metrical properties of no less interest; in particular the special cubical ellipse, which has its imaginary infinity points on the imaginary circle at infinity, possesses remarkable properties to which the cubical hyperbola can offer no parallel.

I am grateful to Mr W. B. Allcock for reading the proofs.

P. W. W.

Aug. 1, 1913.

CONTENTS

NOTATION

Points are in general denoted by the letters P, Q; planes by the letters p, q; lines by the letter λ; dashes and suffixes being added as required.

Taking any tetrahedron of reference, the homogeneous coordinates of any point P are proportional to (x, y, z, t), the absolute values of the coordinates not being stated; the coordinates of P_r are always taken to be proportional to (x_r, y_r, z_r, t_r).

The equation

$$Xx + Yy + Zz + Tt = 0$$

is taken to mean that the point (x, y, z, t) and the plane with coordinates (X, Y, Z, T) are united; the equation of the point P_r is

$$x_r X + y_r Y + z_r Z + t_r T = 0;$$

the coordinates of the plane p_r are (X_r, Y_r, Z_r, T_r) and its equation is

$$X_r x + Y_r y + Z_r z + T_r t = 0.$$

The coordinates of a line λ are taken as (l, m, n, L, M, N) where

$$lL + mM + nN \equiv 0;$$

if P_1, P_2 are any two points on the line, the coordinates are proportional to the second order determinants of the array

$$\begin{vmatrix} x_1 & y_1 & z_1 & t_1 \\ x_2 & y_2 & z_2 & t_2 \end{vmatrix};$$

thus:—

$$\frac{l}{x_1 t_2 - x_2 t_1} = \frac{m}{y_1 t_2 - y_2 t_1} = \frac{n}{z_1 t_2 - z_2 t_1} = \frac{L}{y_1 z_2 - y_2 z_1} = \frac{M}{z_1 x_2 - z_2 x_1} = \frac{N}{x_1 y_2 - x_2 y_1};$$

and, if p_1, p_2 are any two planes through the line, we have correlatively

$$\frac{l}{Y_1 Z_2 - Y_2 Z_1} = \frac{m}{Z_1 X_2 - Z_2 X_1} = \frac{n}{X_1 Y_2 - X_2 Y_1}$$

$$= \frac{L}{X_1 T_2 - X_2 T_1} = \frac{M}{Y_1 T_2 - Y_2 T_1} = \frac{N}{Z_1 T_2 - Z_2 T_1}.$$

Unless the contrary is stated it is to be understood that the line λ_r has coordinates proportional to $(l_r, m_r, n_r, L_r, M_r, N_r)$.

The equations of the faces of the tetrahedron of reference being $x=0$, $y=0$, $z=0$, $t=0$, the faces themselves may be called x, y, z, t; in the same way the vertices may be named X, Y, Z, T, so that the point X is the same as the point yzt, and the plane x is the same as the plane YZT. The six edges of the tetrahedron are named

$$yz, \quad zx, \quad xy, \quad xt, \quad yt, \quad zt,$$

or

$$XT, \quad YT, \quad ZT, \quad YZ, \quad ZX, \quad XY.$$

The following formulae are made use of in the course of the analysis:

(i) The equation of the plane through P_1 and the line λ is

$$(ny_1 - mz_1 + Lt_1)\,x + (lz_1 - nx_1 + Mt_1)\,y + (mx_1 - ly_1 + Nt_1)\,z - (Lx_1 + My_1 + Nz_1)\,t = 0.$$

(ii) The equation of the point in which λ meets the plane p_1 is

$$(NY_1 - MZ_1 + lT_1)\,X + (LZ_1 - NX_1 + mT_1)\,Y + (MX_1 - LY_1 + nT_1)\,Z - (lX_1 + mY_1 + nZ_1)\,T = 0.$$

(iii) The condition of intersection of the two lines λ_1, λ_2 is

$$\varpi_{12} \equiv L_1 l_2 + M_1 m_2 + N_1 n_2 + l_1 L_2 + m_1 M_2 + n_1 N_2 = 0.$$

The cross-ratio of four collinear points P_1, P_2, P_3, P_4 is always taken as

$$(P_1 P_2 P_3 P_4) \equiv P_1P_3 \,.\, P_2P_4 / P_1P_4 \,.\, P_2P_3.$$

I

PROJECTIVE PROPERTIES OF THE TWISTED CUBIC

1. Analytical definition of the twisted cubic.

It is known that if the coordinates of a point P are given by the equations

$$x : y : z : t = a\theta + a' : b\theta + b' : c\theta + c' : d\theta + d',$$

where θ is a parameter, the locus of P is a straight line. The coordinates of the line are

$$(ad' - a'd,\ bd' - b'd,\ cd' - c'd,\ bc' - b'c,\ ca' - c'a,\ ab' - a'b).$$

The parameter θ is homographic; two special points on the line have coordinates (a, b, c, d) and (a', b', c', d'), corresponding to the values ∞, 0 of the parameter.

If the coordinates of a point P are proportional to quadratic functions of a parameter θ, so that

$$x : y : z : t = a_\theta^2 : b_\theta^2 : c_\theta^2 : d_\theta^2,$$

where $a_\theta^2, \ldots$ are written symbolically for $a\theta^2 + 2a'\theta + a'', \ldots$, the locus of P is known to be a conic. The parameter θ is again homographic; the four points corresponding to values $\theta_1, \theta_2, \theta_3, \theta_4$ are projected from any other point of the conic by a pencil, whose cross-ratio is equal to

$$(\theta_1\theta_2\theta_3\theta_4) = (\theta_1 - \theta_3)(\theta_2 - \theta_4)/(\theta_1 - \theta_4)(\theta_2 - \theta_3).$$

Two of the points on the conic (given by $\theta = \infty$, and $\theta = 0$) have coordinates (a, b, c, d) and (a'', b'', c'', d''), and the tangents to the conic at these points meet in the point (a', b', c', d'). The envelope equation of the conic is

$$(aX + bY + cZ + dT)(a''X + b''Y + c''Z + d''T) - (a'X + b'Y + c'Z + d'T)^2 = 0;$$

its locus equations are found by eliminating θ from

$$x : y : z : t = a_\theta^2 : b_\theta^2 : c_\theta^2 : d_\theta^2.$$

From these equations we find the equations of a plane and of some quadric: the plane is unique but the quadric is not unique.

The twisted cubic is defined analytically as the locus of a point whose coordinates are proportional to cubic functions of a single parameter θ, say

$$x : y : z : t = a_\theta^3 : b_\theta^3 : c_\theta^3 : d_\theta^3,$$

where $$a_\theta^3 \equiv a\theta^3 + 3a'\theta^2 + 3a''\theta + a''', \text{ etc.}$$

It is assumed that the four functions a_θ^3, b_θ^3, c_θ^3, d_θ^3 are linearly independent; otherwise the coordinates of any point of the locus would satisfy a linear relation such as

$$Ax + By + Cz + Dt = 0,$$

and the locus would be a plane cubic.

This definition will be taken as the basis of what follows: the properties of the curve will be investigated by analytical processes. The twisted cubic may be defined geometrically in many ways: some of these will be considered in § 26.

An immediate consequence of the definition is that any plane cuts the curve in three points, for the plane p_1 has in common with the curve the points whose parameters are roots of the cubic in θ

$$X_1 a_\theta^3 + Y_1 b_\theta^3 + Z_1 c_\theta^3 + T_1 d_\theta^3 = 0.$$

One of these roots is always real; the other two may be imaginary; if specially two of the roots are equal p_1 contains a tangent line, and if all three roots are equal the plane p_1 is an osculating plane.

2. Equations of the twisted cubic through six given points.

The parametric equations

$$x : y : z : t = a_\theta^3 : b_\theta^3 : c_\theta^3 : d_\theta^3$$

contain the 15 ratios of the 16 constants $a, \ldots, b, \ldots, c, \ldots, d, \ldots$; but the curve is unaltered by any homographic transformation of the parameter θ. Such a homographic transformation contains three arbitrary parameters, and therefore the parametric equations of the curve contain only 12 effective constants. It is therefore to be expected that the curve can be made to satisfy 12 independent conditions and specially can be chosen to pass through any six points. This conclusion, arrived at by counting parameters, will now be verified.

Take four of the points as vertices X, Y, Z, T of the tetrahedron of reference, and let α, β, γ, δ be the values of the parameter at these points. Since $\theta = \alpha$ is to make each of the coordinates y, z, t vanish, $(\theta - \alpha)$ must be a factor of each of the cubics b_θ^3, c_θ^3, d_θ^3; using this argument for each vertex in turn, we have for the parametric equations of the curve

$$x : y : z : t = a/(\theta - \alpha) : b/(\theta - \beta) : c/(\theta - \gamma) : d/(\theta - \delta),$$

where a, b, c, d are constants at present undetermined.

Let the other two points be P_1, P_2, and let the values of the parameter associated with them be 0, ∞ respectively. Then

$$x_1 : y_1 : z_1 : t_1 = a/\alpha : b/\beta : c/\gamma : d/\delta,$$

and $$x_2 : y_2 : z_2 : t_2 = a : b : c : d.$$

Hence $$\alpha : \beta : \gamma : \delta = x_2/x_1 : y_2/y_1 : z_2/z_1 : t_2/t_1,$$

and the parametric equations are

$$x : y : z : t = \frac{x_2}{k\theta - x_2/x_1} : \frac{y_2}{k\theta - y_2/y_1} : \frac{z_2}{k\theta - z_2/z_1} : \frac{t_2}{k\theta - t_2/t_1},$$

where k is an arbitrary constant available for the association of any value of the parameter with any arbitrary point of the curve.

It follows that, by altering the scales of measurement of the coordinates, the parametric equations of a twisted cubic can be reduced to the form

$$x : y : z : t = 1/(\theta - \alpha) : 1/(\theta - \beta) : 1/(\theta - \gamma) : 1/(\theta - \delta),$$

where α, β, γ, δ are the parameters associated with the vertices of the tetrahedron of reference; three of these parameters may be chosen arbitrarily. This form of the equations is specially suitable when we are dealing with four prescribed points of the curve (cf. § 22).

3. Special choice of the tetrahedron of reference.

From the equations

$$x : y : z : t = a_\theta^3 : b_\theta^3 : c_\theta^3 : d_\theta^3,$$

we derive, on solving for θ^3, θ^2, θ,

$$\theta^3 : \theta^2 : \theta : 1 = Ax + By + Cz + Dt : A'x + \dots : A''x + \dots : A'''x + \dots,$$

where A, B, ... are functions of a, ..., b, ..., c, ..., d, ... The solution is always possible since a_θ^3, b_θ^3, c_θ^3, d_θ^3 are by hypothesis linearly independent. If we take the four planes $Ax + By + Cz + Dt = 0$, etc. as the faces of a new tetrahedron of reference, the parametric equations of the cubic may be written*

$$x : y : z : t = \theta^3 : \theta^2 : \theta : \tfrac{1}{3}.$$

It follows that any twisted cubic can be transformed linearly into any other and that all twisted cubics have the same projective properties. Some of these projective properties will be investigated, using this simplified form of the parametric equations.

The tetrahedron of reference is related to the curve in the following way:

(i) the vertices X, T are points of the curve given by $\theta = \infty, 0$.

(ii) the plane $x = 0$ cuts the curve in points given by $\theta^3 = 0$, and is therefore the osculating plane at T.

(iii) the plane $y = 0$ cuts the curve at $X (\theta = \infty)$ and in two points coincident with $T (\theta = 0)$: it is therefore the plane through X and the tangent to the curve at T.

(iv) the plane $z = 0$ is the plane through T and the tangent to the curve at X.

(v) the plane $t = 0$ is the osculating plane at X.

(vi) the vertex Y is the point in which the tangent at X meets the osculating plane at T.

(vii) the vertex Z is the point in which the tangent at T meets the osculating plane at X.

The curve is the locus of intersections of corresponding planes of the three homographic pencils given by

$$x - \theta y = 0, \quad y - \theta z = 0, \quad z - 3\theta t = 0,$$

the axes of the pencils being the tangents at X and T and the chord XT.

4. Quadrics containing the cubic.

The quadric

$$s \equiv (a, b, c, d, f, g, h, u, v, w \between x, y, z, t)^2 = 0$$

contains the point $(\theta^3, \theta^2, \theta, \tfrac{1}{3})$ for all values of θ, if

$$(a, b, c, d, f, g, h, u, v, w \between \theta^3, \theta^2, \theta, \tfrac{1}{3})^2 \equiv 0.$$

* It is usual to take the parametric equations as $x : y : z : t = \theta^3 : \theta^2 : \theta : 1$; the reason for introducing the coefficient 3 will appear in § 9.

This requires

$$a=0,\quad h=0,\quad b+2g=0,\quad u+3f=0,\quad 3c+2v=0,\quad w=0,\quad d=0,$$

and therefore the equation of any such quadric can be written in the form

$$A(zx-y^2)+B(3xt-yz)+C(3yt-z^2)=0,$$

where A, B, C are constants.

It follows that

(i) a unique quadric can be drawn through the curve and any two arbitrary points.

(ii) if s, s', s'' are any three linearly independent quadrics through the curve, any other quadric through the curve is given by the equation

$$ks+k's'+k''s''=0.$$

Any quadric containing the twisted cubic will be spoken of as a "circumscribing quadric."

5. Quadric cones circumscribing the cubic.

If the equation

$$A(zx-y^2)+B(3xt-yz)+C(3yt-z^2)=0$$

represents a cone, the coordinates of its vertex are given by the equations

$$Az+3Bt=0,\quad -2Ay-Bz+3Ct=0,\quad Ax-By-2Cz=0,\quad Bx+Cy=0.$$

The consistency of these four equations requires $B^2-AC=0$, and the coordinates of the vertex are then given by the equations

$$x:y:z:3t=BC:-B^2:AB:-A^2.$$

Writing $B/A=-\theta$, $C/A=\theta^2$, the coordinates of the vertex are given by

$$x:y:z:t=\theta^3:\theta^2:\theta:\tfrac{1}{3}.$$

Hence any quadric cone circumscribing the curve has its vertex on the curve.

In general, if the curve is projected by a cone from any point in space, a plane through the vertex cuts the curve in three points, and the projecting cone in three generators, so that the cone is a cubic cone. The equation of such a cubic cone is given in § 15. The present result shows that if the vertex of the cone lies on the curve the projecting cone degenerates into a quadric cone.

The equation of the quadric cone projecting the curve from the point with parameter θ is found from the values of $A : B : C$ above to be

$$(zx - y^2) - \theta (3xt - yz) + \theta^2 (3yt - z^2) = 0.$$

The equation of the tangent plane to this cone along the generator through the point of the curve with parameter ϕ is found to be

$$x - (\theta + 2\phi) y + \phi (\phi + 2\theta) z - 3\theta\phi^2 t = 0.$$

This plane cuts the curve in points whose parameters θ' are given by $(\theta' - \phi)^2 (\theta' - \theta) = 0$, and therefore the plane touches the curve at the point whose parameter is ϕ (see § 6); if we put $\theta = \phi$, the equation

$$x/3 - \theta y + \theta^2 z - \theta^3 t = 0$$

gives the tangent plane to the cone along the generator which is the tangent to the curve at the vertex. It will be verified in § 7 that this plane is the osculating plane at the point θ.

The first theorem above is sometimes* stated conversely in the form: "the locus of the vertices of quadric cones through six given points is the unique twisted cubic passing through those points." This is not true; for if s, s', s'', s''' are four quadrics through the six points any quadric through the six points has an equation of the form

$$\kappa s + \kappa' s' + \kappa'' s'' + \kappa''' s''' = 0,$$

and therefore the locus of the vertices of the quadric cones is given by equating to zero the Jacobian of s, s', s'', s'''. It is a quartic surface† with 25 lines on it, 15 of which join pairs of the six points and the remaining 10 are the intersections of planes through any three of the points and the other three points; the quartic surface passes through the unique twisted cubic determined by the six points.

The curve is projected from any two points P_1, P_2 of itself by two quadric cones: the quartic curve of intersection of these two cones is made up of the line P_1P_2 and the twisted cubic itself.

6. The cross-ratio of four points on the curve.

The plane

$$p_1 \equiv X_1 x + Y_1 y + Z_1 z + T_1 t = 0$$

cuts the curve in three points whose parameters are given by

$$X_1 \theta^3 + Y_1 \theta^2 + Z_1 \theta + T_1/3 = 0.$$

* See Chasles, *Aperçu Historique*, note XXXIII, § 2.

† Salmon, *Geometry of Three Dimensions*, § 571.

Calling these parameters α, β, γ, we have

$$X_1 : Y_1 : Z_1 : T_1/3 = 1 : -\Sigma\alpha : \Sigma\beta\gamma : -\alpha\beta\gamma,$$

and therefore the equation of the plane through the three points with parameters α, β, γ is

$$x - y\Sigma\alpha + z\Sigma\beta\gamma - 3t\alpha\beta\gamma = 0.$$

If the points α, β are fixed, while γ is variable, we get a pencil of planes determined by the homographic parameter γ; the cross-ratio of the four planes through the chord $\alpha\beta$ and the points γ_1, γ_2, γ_3, γ_4 is equal to

$$(\gamma_1\gamma_2\gamma_3\gamma_4) = (\gamma_1 - \gamma_3)(\gamma_2 - \gamma_4)/(\gamma_1 - \gamma_4)(\gamma_2 - \gamma_3),$$

and is independent of the chord $\alpha\beta$. We can therefore speak of the cross-ratio of four points on the curve, the parameters determining the points of the cubic being homographic. It was proved in § 5 that the cubic is projected from any one of its points by a quadric cone; associated with any four generators of a quadric cone is the cross-ratio of the pencil of planes through them and a variable generator. The cross-ratio of four points of the cubic is also the cross-ratio of the four generators, through these points, of any circumscribing quadric cone.

Consider the tetrahedron with its vertices at the points of the cubic given by the parameters γ_1, γ_2, γ_3, γ_4; the pencil of planes through the chord $\alpha\beta$ and the vertices of this tetrahedron have a cross-ratio $(\gamma_1\,\gamma_2\,\gamma_3\,\gamma_4)$ which is independent of the chord $\alpha\beta$. All chords of the cubic are therefore rays of the tetrahedral complex of the tetrahedron whose vertices are at any four arbitrary points γ_1, γ_2, γ_3, γ_4 of the curve, and the cross-ratio associated with the tetrahedral complex is $(\gamma_1\,\gamma_2\,\gamma_3\,\gamma_4)$. The tangents of the curve are rays of any such complex. This aspect will be discussed further in §§ 16, 17.

7. The osculating plane at any point of the curve.

The plane through the points with parameters α, β, γ of the curve is given by the equation

$$x - y\Sigma\alpha + z\Sigma\beta\gamma - 3t\alpha\beta\gamma = 0.$$

If these three points are all coincident with the point with parameter θ, we get

$$x/3 - \theta y + \theta^2 z - \theta^3 t = 0$$

as the equation of the osculating plane at this point. The coordinates of any osculating plane are given parametrically by

$$X : Y : Z : T = \tfrac{1}{3} : -\theta : \theta^2 : -\theta^3.$$

Three osculating planes pass through any point P_0, the parameters θ_1, θ_2, θ_3 of the points of osculation being the roots of the cubic

$$t_0\theta^3 - z_0\theta^2 + y_0\theta - x_0/3 = 0.$$

The plane through these three points is given (§ 6) by the equation

$$x - y\Sigma\theta_1 + z\Sigma\theta_2\theta_3 - 3t\theta_1\theta_2\theta_3 = 0,$$

or

$$t_0 x - z_0 y + y_0 z - x_0 t = 0,$$

and therefore contains the point P_0.

Hence any three osculating planes meet in a point coplanar with their points of osculation.

8. The tangent at any point.

The coordinates of the tangent at the point θ are proportional to the second order determinants formed from the array

$$\begin{vmatrix} \theta^3 & \theta^2 & \theta & \frac{1}{3} \\ 3\theta^2 & 2\theta & 1 & . \end{vmatrix},$$

and are given by

$$(\theta^2, \ \tfrac{2}{3}\theta, \ \tfrac{1}{3}, \ \theta^2, \ -2\theta^3, \ \theta^4).$$

It follows that

(i) any line in space is met by four tangents to the curve;

(ii) the tangents are rays of the screw (linear complex) whose equation is

$$l - L = 0;$$

(iii) the tangents are also rays of the tetrahedral complex

$$\frac{lL}{3} = -\frac{mM}{4} = nN.$$

The second of these results is of fundamental importance: the only linear relation connecting the coordinates of a tangent is $l - L = 0$, so that there is a unique screw having the tangents of the cubic included among its rays. We shall call $l - L = 0$ "the screw containing the tangents."

It is worthy of remark that the vertices of the tetrahedron of reference do not all lie on the curve, so that the third result is not a special case of the result of § 6.

9. The screw containing the tangents.

Since the equation of the screw is $l - L = 0$, the equation of the polar plane of any point P_1 is

$$t_1 x - z_1 y + y_1 z - x_1 t = 0.$$

This plane contains the three points whose osculating planes pass through P_1 (§ 7); in other words the pole of any plane is the common point of the three osculating planes at the points where the plane meets the curve.

Specially any point of the curve and the osculating plane thereat are pole and polar with respect to the screw.

Polarization with respect to the screw is effected analytically by the transformation from point coordinates to plane coordinates given by the equations

$$x : y : z : t = T : -Z : Y : -X,$$

or equivalently $\quad X : Y : Z : T = t : -z : y : -x.$

The object of choosing the parametric equations in the form

$$x : y : z : t = \theta^3 : \theta^2 : \theta : \tfrac{1}{3},$$

rather than the more usual form

$$x : y : z : t = \theta^3 : \theta^2 : \theta : 1,$$

is to make the equations of the polarizing transformation symmetrical as between point coordinates and plane coordinates.

The polar line of (l, m, n, L, M, N) with respect to the screw has coordinates (L, m, n, l, M, N).

Any property of the twisted cubic leads to a correlative property: so far the curve has been regarded as a locus of points. Its analytical definition (§ 1) may be put in the equivalent form "a twisted cubic is a curve osculated by planes whose coordinates are proportional to cubic functions of a parameter."

A quadric touching all the osculating planes of the curve is said to be "inscribed" in the curve; from §§ 4—6 we have the results

(1) any quadric inscribed in the twisted cubic has for its equation

$$A(YT - Z^2) + B(YZ - 3XT) + C(3ZX - Y^2) = 0.$$

(2) if a conic is inscribed in a twisted cubic, its plane osculates the curve.

(3) the inscribed conic lying in the osculating plane at the point θ is given by the equation

$$(YT - Z^2) - \theta(YZ - 3XT) + \theta^2(3ZX - Y^2) = 0.$$

(4) the correlative of a chord of the curve is a "line in two planes," that is a line of intersection of two osculating planes; a line-in-two-planes is the polar line of a chord with respect to the screw containing the tangents. Four fixed osculating planes cut a variable

line-in-two-planes in four points of constant cross-ratio, which may be called the cross-ratio of the four osculating planes, and this cross-ratio is equal to that of the four points of osculation as defined in § 6. Also from (2) any osculating plane has in it an inscribed conic touching all the osculating planes of the curve; any four tangents of this conic have an associated cross-ratio which is equal to the cross-ratio of the four corresponding osculating planes.

(5) it follows also from § 6 that all lines-in-two-planes are rays of a tetrahedral complex whose tetrahedron has for its faces any four osculating planes of the curve.

10. Chords of the curve as generators of a circumscribing quadric.

The quadric given by

$$A(zx-y^2)+B(3xt-yz)+C(3yt-z^2)=0$$

circumscribes the cubic for all values of A, B, C; the ends of the chord joining the points with parameters θ, ϕ lie on the quadric, and the chord is therefore a generator of the quadric if a third point, say

$$(\theta^3-\phi^3,\ \theta^2-\phi^2,\ \theta-\phi,\ 0),$$

or

$$(\theta^2+\theta\phi+\phi^2,\ \theta+\phi,\ 1,\ 0),$$

also lies on the quadric. The condition for this reduces to

$$A\theta\phi+B(\theta+\phi)+C=0;$$

hence associated with any circumscribing quadric there is an involution of points on the cubic such that chords joining corresponding points are generators of the quadric; clearly the generators must belong to the same system, for two intersecting chords would lie in a plane cutting the curve in four points. Each of the generators of the other system will meet the curve once. Included among the generators are the tangents to the curve at the double points of the involution, with parameters given by the equation

$$A\theta^2+2B\theta+C=0.$$

Two such circumscribing quadrics given by

$$A(zx-y^2)+B(3xt-yz)+C(3yt-z^2)=0,$$

$$A'(zx-y^2)+B'(3xt-yz)+C'(3yt-z^2)=0,$$

have as a common generator the chord joining the points with parameters θ, ϕ given by the equations

$$A\theta\phi + B(\theta+\phi) + C = 0,$$

and
$$A'\theta\phi + B'(\theta+\phi) + C' = 0;$$

the parameters of the ends of the chord are the roots of the quadratic

$$\begin{vmatrix} \theta^2 & -\theta & 1 \\ C & B & A \\ C' & B' & A' \end{vmatrix} = 0.$$

If the circumscribing quadric is a cone whose vertex is at the point of the curve with parameter θ_0, the corresponding involution is degenerate, since all the points of the curve are paired with θ_0. This requires $(\theta-\theta_0)$ to be a factor of $A\theta\phi + B(\theta+\phi) + C$, and so

$$A\theta_0 + B = 0, \quad B\theta_0 + C = 0:$$

these equations lead to the equation of the quadric cone projecting the curve from θ_0, namely

$$\theta_0^2(3yt - z^2) - \theta_0(3xt - yz) + (zx - y^2) = 0.$$

This last equation (found otherwise in § 5) expresses the condition that the point P and the point θ_0 of the curve are collinear with some other point of the curve. Hence regarding P as fixed the equation gives the parameters (θ_0) of the ends of the unique chord* through P.

The quadratic equation in θ

$$(3yt - z^2)\theta^2 - (3xt - yz)\theta + (zx - y^2) = 0$$

is the Hessian of the cubic equation

$$t\theta^3 - z\theta^2 + y\theta - x/3 = 0,$$

which gives the parameters of the points P_1, P_2, P_3, whose osculating planes pass through P. If the ends of the chord are real, then two of the points P_1, P_2, P_3 are imaginary, and conversely; if the ends of the chord are imaginary, the points P_1, P_2, P_3 are real, and conversely. In other words the polar plane of any point on a chord with real ends cuts the curve in only one real point, and the unique chord through the pole of any plane cutting the curve in three real points has imaginary ends.

* If the twisted cubic is projected from P on to any plane, the resulting curve will be a plane cubic with a node at the point where the unique chord through P cuts the plane.

11. Lines-in-two-planes as generators of an inscribed quadric.

Polarization of the results of § 10 with respect to the screw containing the tangents leads to a series of reciprocal theorems, wherein chords are replaced by lines common to two osculating planes.

(i) Associated with any inscribed quadric (§ 9) is an involution of osculating planes of the curve with the property that the line of intersection of any two companion osculating planes is a generator of the quadric.

(ii) In any plane there is a unique line which is the intersection of two osculating planes.

Take any point P and its polar plane p with respect to the screw containing the tangents: through P passes a unique chord whose ends are Q, Q': if the osculating planes at these points are q and q', the line qq' is the unique line-in-two-planes of the plane p. The two

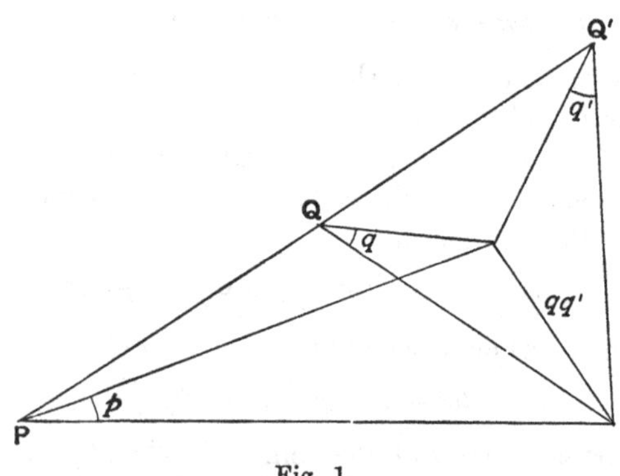

Fig. 1.

lines QQ', qq' are polar lines with respect to the screw. The points Q, Q' (and the planes q, q') are both imaginary or both real according as p cuts the curve in three real points, or in one real and two imaginary points.

If specially we take the plane p to be the plane at infinity, the line qq' will be the intersection of two parallel osculating planes: a twisted cubic has therefore a unique pair of parallel osculating planes, which are real only when the curve has only one real point at infinity (see § 44, (iii)).

12. The harmonic polar of any plane with respect to the three osculating planes through its pole is the unique chord through that pole.

It may be well to recall first the definition of a harmonic pole and polar with respect to any triangle. Take any triangle $P_1P_2P_3$ and a coplanar point P_4, and let P_1P_4, P_2P_4, P_3P_4 cut the opposite sides of the triangle in points Q_1, Q_2, Q_3. If P_2P_3 meets Q_2Q_3 in the point

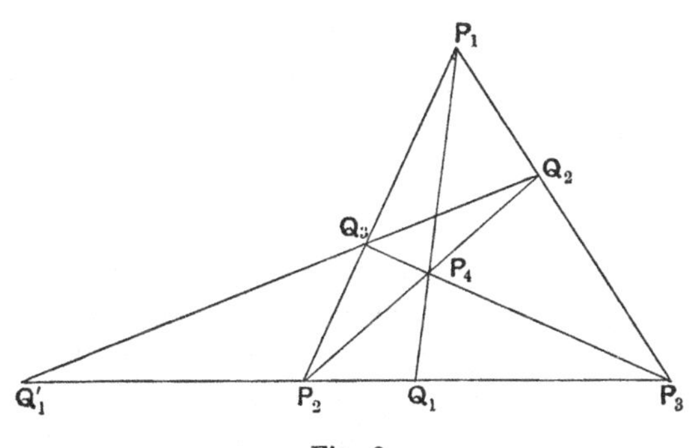

Fig. 2.

Q_1', then Q_1 and Q_1' are harmonically separated by P_2, P_3 and the three such points Q_1', Q_2', Q_3' are collinear. The point P_4 and the line $Q_1'Q_2'Q_3'$ are said to be harmonic pole and polar with regard to the triangle*.

The analytical definition may be stated concisely thus: if the equations of any four coplanar points P_1, P_2, P_3, P_4 in space are

$$P_1 \equiv x_1X + y_1Y + z_1Z + t_1T = 0,$$
$$P_2 \equiv x_2X + y_2Y + z_2Z + t_2T = 0,$$
$$P_3 \equiv x_3X + y_3Y + z_3Z + t_3T = 0,$$

and
$$P_4 \equiv x_4X + y_4Y + z_4Z + t_4T = 0,$$

then there is an identical relation of the form

$$k_1P_1 + k_2P_2 + k_3P_3 + k_4P_4 \equiv 0,$$

where k_1, k_2, k_3, k_4 are constants; and the equations of the harmonic polar of P_4 with regard to the triangle $P_1P_2P_3$ are

$$k_1P_1 = k_2P_2 = k_3P_3.$$

* If the line is at infinity its harmonic pole is the centroid of the triangle.

To identify this with the geometrical definition:—by virtue of the identity, each of the equations

$$k_2P_2 + k_3P_3 = 0, \quad k_1P_1 + k_4P_4 = 0$$

represents the same point. This point is collinear with P_2, P_3 and also with P_1, P_4 and must therefore be the point Q_1. The point Q_1' is the harmonic conjugate of Q_1 with respect to P_2, P_3 and therefore its equation is

$$k_2P_2 - k_3P_3 = 0.$$

The points Q_2', Q_3' are given similarly by the equations

$$k_3P_3 - k_1P_1 = 0, \quad k_1P_1 - k_2P_2 = 0,$$

and therefore the three points Q_1', Q_2', Q_3' lie on the line given by

$$k_1P_1 = k_2P_2 = k_3P_3.$$

If we reciprocate the plane figure with regard to any quadric (or project it from any point in space), we get a figure consisting of three given planes, a plane through their common point, and a line which is defined to be the harmonic polar of the last plane with regard to the first three. The analytical correlative is:

If $p_1 = 0$, $p_2 = 0$, $p_3 = 0$, $p_4 = 0$ are the equations of four concurrent planes, then there is an identity

$$k_1p_1 + k_2p_2 + k_3p_3 + k_4p_4 \equiv 0,$$

and the equations of the harmonic polar line of p_4 with regard to p_1, p_2, p_3 are

$$k_1p_1 = k_2p_2 = k_3p_3.$$

The section of the solid figure by any plane is a triangle with a point and line which are harmonic pole and polar with respect to the triangle.

Now consider the plane through the points of the cubic with parameters θ_1, θ_2, θ_3; its equation is (§ 6)

$$p \equiv x - y\Sigma\theta_1 + z\Sigma\theta_2\theta_3 - 3t\theta_1\theta_2\theta_3 = 0.$$

The equations of the osculating planes at the three points are

$$p_1 \equiv x/3 - \theta_1 y + \theta_1^2 z - \theta_1^3 t = 0,$$
$$p_2 \equiv x/3 - \theta_2 y + \theta_2^2 z - \theta_2^3 t = 0,$$
$$p_3 \equiv x/3 - \theta_3 y + \theta_3^2 z - \theta_3^3 t = 0.$$

The four planes are concurrent in the pole of p with respect to the screw containing the tangents, and the identical relation between the equations of the four concurrent planes is

$$(\theta_2 - \theta_3)^3 p_1 + (\theta_3 - \theta_1)^3 p_2 + (\theta_1 - \theta_2)^3 p_3 \equiv (\theta_2 - \theta_3)(\theta_3 - \theta_1)(\theta_1 - \theta_2)\, p.$$

The equations of the harmonic polar of the plane p with regard to the planes p_1, p_2, p_3 are therefore

$$(\theta_2-\theta_3)^3 p_1 = (\theta_3-\theta_1)^3 p_2 = (\theta_1-\theta_2)^3 p_3.$$

This harmonic polar will meet the curve in a point with parameter ϕ, if ϕ can be found to satisfy the equations

$$(\theta_2-\theta_3)^3(\phi-\theta_1)^3 = (\theta_3-\theta_1)^3(\phi-\theta_2)^3 = (\theta_1-\theta_2)^3(\phi-\theta_3)^3.$$

These equations are equivalent to the single condition that the cross-ratio $(\theta_1\,\theta_2\,\theta_3\,\phi)$ should have one of the values $-\omega$, $-\omega^2$, where ω is an imaginary cube root of unity, and this condition is satisfied if ϕ is either root of the Hessian of the cubic whose roots are θ_1, θ_2, θ_3. In § 10 it was proved that these two values of ϕ determine the ends of the unique chord through the pole of p; the result is therefore established.

The correlative theorem may be stated: the pole of any plane p and the unique line-in-two-planes lying in p are harmonic pole and polar with respect to the three points in which p cuts the cubic.

13. The developable of tangents.

The parameters of the ends of the chord through the point P are (§ 10) given by

$$(3yt-z^2)\,\theta^2-(3xt-yz)\,\theta+(zx-y^2)=0\,;$$

if the ends of the chord are coincident P lies on a tangent to the curve, and therefore the equation of the surface generated by the tangents is

$$(3xt-yz)^2-4\,(zx-y^2)\,(3yt-z^2)=0.$$

The same equation is found by expressing the condition that the cubic in θ

$$t\theta^3-z\theta^2+y\theta-x/3=0$$

should have equal roots, for in this case two of the osculating planes through P are coincident.

The surface is a developable, its tangent planes being the osculating planes of the curve: the twisted cubic is the edge of regression (cuspidal edge) of the developable. Any line cuts the surface in four points through each of which passes a tangent to the curve.

Any plane cuts the surface in a quartic curve, which has three cusps at the points where the plane cuts the cubic: the section by a plane through a tangent to the curve (i.e. through a generator of the developable) consists of this tangent and a plane cubic with a cusp at the remaining point of intersection of the plane with the twisted cubic: the section by an osculating plane is a conic and a pair of lines coincident with the tangent at the point of osculation (this conic has already (§ 9) been noticed as a conic inscribed in the twisted cubic).

The correlative equation

$$(3XT - YZ)^2 - 4(3ZX - Y^2)(YT - Z^2) = 0$$

is the envelope equation of the twisted cubic: since it is of the fourth degree, four tangent planes to the cubic pass through any line.

The relations between the cubic curve and the developable of tangents may be summarized thus:

(i) If the points of the cubic are polarized with respect to the screw containing the tangents, we get the osculating planes of the cubic, and conversely.

(ii) The points lying on the tangents to the curve lie on the developable, which is a locus of the fourth order; the planes through the tangents to the curve envelope the curve itself, which is an envelope of the fourth class. If the developable is polarized with respect to the screw containing the tangents, we get the curve itself, and conversely.

(iii) The curve is a locus of the third order, and the developable is of the third class. A circumscribing quadric contains all the points of the curve, and an inscribed quadric touches all the planes of the developable.

14. A property of two harmonic pairs of points on the curve.

It will now be proved that, if the points P, Q are separated harmonically by the points A, B, then the tangent at P meets the osculating plane at Q in a point lying on the plane QAB.

Let P, Q be taken as the vertices X, T of the tetrahedron of reference chosen in § 3, then the tangent at X meets the osculating plane at T in the point Y. If α, β are the parameters of the points A, B, the equation of the plane TAB is

$$x - (\alpha + \beta) y + \alpha\beta z = 0,$$

and this plane passes through Y if $\alpha + \beta = 0$. Also $\alpha + \beta = 0$ is the condition that the points A, B are harmonically separated by the points X, T with parameters ∞, 0.

An application of this result to the case of the cubical hyperbola is made in § 34.

15. Conditions that a line should (i) meet the cubic, (ii) lie in an osculating plane.

The line λ with coordinates (l, m, n, L, M, N) is the intersection of any two of the planes given by the equations

$$\begin{aligned} ny - mz + Lt &= 0, \\ lz - nx + Mt &= 0, \\ mx - ly + Nt &= 0, \\ Lx + My + Nz &= 0. \end{aligned}$$

The line will meet the curve in the point with parameter θ, if θ can be found to satisfy simultaneously any two of the four equations

$$n\theta^2 - m\theta + \tfrac{1}{3}L = 0 \quad \text{(i)},$$
$$l\theta - n\theta^3 + \tfrac{1}{3}M = 0 \quad \text{(ii)},$$
$$m\theta^3 - l\theta^2 + \tfrac{1}{3}N = 0 \quad \text{(iii)},$$
$$L\theta^2 + M\theta + N = 0 \quad \text{(iv)}.$$

From (i) and (iv), we have

$$\theta^2 : \theta : 1 = LM + 3mN : 3nN - L^2 : -3(mL + nM),$$

and, using the relation $lL + mM + nN = 0$, (ii) gives

$$-3(mL + nM)\,\theta^3 = (3lN - M^2 + NL),$$

and (iii) gives

$$-(mL + nM)/\theta = 3nl - 3m^2 + nL.$$

So the line meets the cubic in a point with parameter θ, if

$$\theta^4 : \theta^3 : \theta^2 : \theta : 1 = A : B : C : D : E \quad \text{(v)},$$

where $A \equiv 3lN - M^2 + NL$, $B \equiv LM + 3mN$, $C \equiv 3nN - L^2$,

$$D \equiv -3(mL + nM), \quad E \equiv 9(nl - m^2 + \tfrac{1}{3}nL).$$

The condition of intersection is found by eliminating θ; writing

$$\Pi \equiv \tfrac{1}{3}L^3 - lL^2 - 3(l + L)\,nN + 3m^2N + nM^2,$$

we find

$$AC - B^2 \equiv -3N\Pi, \quad AD - BC \equiv 3M\Pi, \quad AE - BD \equiv -9l\Pi,$$
$$BD - C^2 \equiv -3L\Pi, \quad BE - CD \equiv -9m\Pi, \quad CE - D^2 \equiv -9n\Pi,$$

and the required condition is $\Pi = 0$.

Lines meeting the curve are therefore rays of a cubic complex; the equation of the cone projecting the curve from any point P_1 is found by substituting the values

$$t_1x - x_1t, \quad t_1y - y_1t, \quad t_1z - z_1t, \quad z_1y - y_1z, \quad x_1z - z_1x, \quad y_1x - x_1y$$

for l, m, n, L, M, N respectively in the preceding equation $\Pi = 0$.

Correlatively, if the line λ lies in an osculating plane, its polar line with respect to the screw containing the tangents, with coordinates (L, m, n, l, M, N), meets the cubic, and therefore the condition is

$$\Pi' \equiv l^3/3 - l^2L - 3(L+l)Nn + 3m^2N + nM^2 = 0.$$

The aggregate of such lines makes up a cubic complex; the equation of the class-cubic curve in any plane p_1, which is the envelope of all the osculating planes of the twisted cubic, is found by substituting the values

$$Y_1Z - Z_1Y, \quad Z_1X - X_1Z, \quad X_1Y - Y_1X,$$
$$T_1X - X_1T, \quad T_1Y - Y_1T, \quad T_1Z - Z_1T$$

for l, m, n, L, M, N respectively in the equation $\Pi' = 0$.

We have already seen (§ 9) that this cubic curve degenerates into a conic when p_1 is an osculating plane of the twisted cubic.

The identity $\qquad \Pi - \Pi' \equiv \frac{1}{3}(L-l)^3$

verifies that a line which lies in an osculating plane and also meets the cubic is a ray of the screw containing the tangents.

16. Conditions that a line should be a chord of the cubic.

If λ is a chord of the cubic, the equations (i)—(iv) of § 15 are satisfied simultaneously by the same two values of θ, and the equations (v) are indeterminate. The chords of the cubic are therefore rays of the five quadratic complexes whose equations are

$$A = 0, \; B = 0, \; C = 0, \; D = 0, \; E = 0.$$

These five equations are linearly independent, but are algebraically equivalent to two only.

If we write down the equation of the most general quadratic complex with its 21 terms, we can find the condition that it may contain all the chords of the curve by substituting

$$l = \frac{M^2 - NL}{3N}, \quad m = -\frac{LM}{3N}, \quad n = \frac{L^2}{3N},$$

and making the resulting quadratic in L, M, N an identical zero. If this is done, it will be found that the equation of the complex is a linear multiple of the five equations above, and can therefore be written in the form $aA + bB + cC + dD + eE = 0$, where a, b, c, d, e are arbitrary constants. It follows that there are ∞^4 quadratic complexes containing the chords of the cubic.

17. Tetrahedral complexes containing the chords of the cubic.

We have seen already (§ 6) that, if four fixed points are taken on the curve, the planes through them and any variable chord form a pencil whose cross-ratio is equal to that of the four fixed points. The chords of the curve must therefore belong to the tetrahedral complex associated with the four fixed points and their cross-ratio. Any such tetrahedron can be chosen in ∞^4 ways, and therefore the chords of the cubic belong to ∞^4 tetrahedral complexes. Taking this in connection with the result of § 16, it is natural to suppose that every quadratic complex containing the chords is a tetrahedral complex.

Any quadratic complex containing all the chords of the cubic has an equation of the form $aA + bB + cC + dD + eE = 0$, and, referring to the equations (v) of § 15, we see that every line meeting the cubic at the point whose parameter is θ is a ray of this complex if

$$a\theta^4 + b\theta^3 + c\theta^2 + d\theta + e = 0.$$

This equation determines four points on the cubic each with the property that all lines through it are rays of the complex. The quadratic complex is therefore the tetrahedral complex associated with the tetrahedron which has its vertices at the four points.

The same result can be obtained less simply as follows without using the formulae of § 15:

Taking any four points P_1, P_2, P_3, P_4 and a line λ it is found that the cross-ratio of the pencil of planes $\lambda(P_1, P_2, P_3, P_4)$ has the value

$$[13][24]/[14][23],$$

where

$$[13] \equiv l(y_1z_3 - y_3z_1) + m(z_1x_3 - z_3x_1) + n(x_1y_3 - x_3y_1) + L(x_1t_3 - x_3t_1) + M(y_1t_3 - y_3t_1) + N(z_1t_3 - z_3t_1),$$

with similar values for [24], [14], [23].

The equation of the tetrahedral complex associated with this tetrahedron and the cross-ratio μ is therefore

$$[13][24] - \mu[14][23] = 0.$$

Now take P_1, P_2, P_3, P_4 as points on the curve with parameters θ_1, θ_2, θ_3, θ_4 and write

$$\mu = (\theta_1 - \theta_3)(\theta_2 - \theta_4)/(\theta_1 - \theta_4)(\theta_2 - \theta_3).$$

After some algebraical reductions it is found that the equation of the tetrahedral complex can be written in the form

$$A - B\Sigma\theta_1 + C\Sigma\theta_1\theta_2 - D\Sigma\theta_1\theta_2\theta_3 + E\theta_1\theta_2\theta_3\theta_4 = 0,$$

where A, B, C, D, E are the same expressions as before.

Included among the ∞^4 tetrahedral complexes are those obtained by taking four of the parameters a, b, c, d, e zero. The tetrahedron associated with the complex $A \equiv 3lN - M^2 + NL = 0$ has its four vertices coincident with the point T, given by $\theta = 0$; the tetrahedron associated with the complex $B \equiv LM + 3mN = 0$ has three vertices coincident with T ($\theta = 0$) and the remaining vertex at X ($\theta = \infty$); the tetrahedron associated with the complex $C \equiv 3nN - L^2 = 0$ has two vertices coincident with X ($\theta = \infty$) and two coincident with T ($\theta = 0$); and the tetrahedra associated with the complexes $D = 0$, $E = 0$ are determined similarly.

The tetrahedral complex given by the equation

$$(A, B, C, D, E \between 1, -\theta)^4 = 0$$

has the four vertices of its associated tetrahedron coincident with the point whose parameter is θ; this complex will be called "the osculating tetrahedral complex of chords at the point θ."

Interchange of l and L in the formulae preceding will give the correlative results concerning the lines-in-two-planes. For instance the tetrahedral complex of lines-in-two-planes given by the equation

$$(A', B', C', D', E' \between 1, -\theta)^4 = 0,$$

where A', B', ... are derived from A, B, ... by interchanging l and L, so that

$$A' \equiv 3LN - M^2 + lN, \text{ etc.},$$

has the faces of its tetrahedron coincident with the osculating plane at the point θ, and may be called "the osculating tetrahedral complex of lines-in-two-planes at the point θ."

18. Twisted cubics touching four given lines.

We have seen that a twisted cubic is determined by 12 parameters; in order that the curve may touch a given line three conditions are necessary, and so the mere counting of parameters suggests that a finite number of cubics can be drawn having four given tangents. This however is not the case.

Consider the tangents at four points of the curve given by

$$x : y : z : t = \theta^3 : \theta^2 : \theta : \tfrac{1}{3}.$$

If λ_1, λ_2, λ_3, λ_4 are the tangents at the points with parameters θ_1, θ_2, θ_3, θ_4, their coordinates are given by the sets of equations

$$\frac{l_1}{\theta_1^2} = \frac{m_1}{\frac{2}{3}\theta_1} = \frac{n_1}{\frac{1}{3}} = \frac{L_1}{\theta_1^2} = \frac{M_1}{-2\theta_1^3} = \frac{N_1}{\theta_1^4}; \text{ etc.}$$

Writing

$$\varpi_{12} \equiv l_1 L_2 + m_1 M_2 + n_1 N_2 + l_2 L_1 + m_2 M_1 + n_2 N_1,$$

so that $\varpi_{12}=0$ is the condition of intersection of the lines λ_1 and λ_2, we have

$$\varpi_{12}=2\theta_1^2\theta_2^2-\tfrac{4}{3}(\theta_1^3\theta_2+\theta_1\theta_2^3)+\tfrac{1}{3}(\theta_1^4+\theta_2^4)$$
$$=\tfrac{1}{3}(\theta_1-\theta_2)^4,$$

with similar values for ϖ_{13}, ϖ_{14}, etc.

Hence, from the identity

$$(\theta_2-\theta_3)(\theta_1-\theta_4)+(\theta_3-\theta_1)(\theta_2-\theta_4)+(\theta_1-\theta_2)(\theta_3-\theta_4)\equiv 0,$$

we have $$(\varpi_{23}\varpi_{14})^{\frac{1}{4}}+(\varpi_{31}\varpi_{24})^{\frac{1}{4}}+(\varpi_{12}\varpi_{34})^{\frac{1}{4}}=0$$

as a condition to be satisfied by the coordinates of any four tangents to a twisted cubic. It follows therefore that four lines cannot be chosen arbitrarily as tangents to a twisted cubic.

The problem is poristic: if the four lines are tangents to any twisted cubic, then ∞^1 cubics can be found having them as tangents. For a full discussion of the question the reader is referred to A. C. Dixon, *Quarterly Journal of Mathematics*, 1889, XXIII; the same paper contains an investigation of twisted cubics passing through five given points.

19. The two transversals of four tangents to the cubic.

Any line λ in space is (see § 8 (i), § 13) met by four tangents: the other transversal of these four tangents is the polar line of λ with respect to the screw containing the tangents, for the parameters of the four contact points are roots of the equation

$$f(\theta)\equiv n\theta^4-2m\theta^3+(l+L)\theta^2+\tfrac{2}{3}M\theta+\tfrac{1}{3}N=0,$$

and the coordinates of the polar line of λ are found by interchanging l and L, so that the equation for the parameters of the contact points is unaltered.

The invariants of this quartic have a simple interpretation: using for a moment the ordinary notation, the quartic

$$(a,\ b,\ c,\ d,\ e\between\theta,\ 1)^4$$

has the two invariants

$$I\equiv ae-4bd+3c^2,$$

$$J\equiv\begin{vmatrix} a & b & c\\ b & c & d\\ c & d & e\end{vmatrix},$$

and the discriminant $$I^3-27J^2.$$

For the quartic $f(\theta)$, we calculate

$$I \equiv \tfrac{1}{12}(l-L)^2,$$
$$-36J \equiv \Pi + \tfrac{1}{6}(l-L)^3$$
$$\equiv \Pi' - \tfrac{1}{6}(l-L)^3,$$

where Π and Π' are the same expressions as in § 15.

The vanishing of I, which requires $l-L=0$, makes the cross-ratio of the four contact points equal to $-\omega$ or $-\omega^2$, where ω is an imaginary cube root of unity. The line λ is a ray of the screw containing the tangents, and the two transversals of the four tangents are coincident. Reference to § 12 shows that if three of the contact points are given the fourth point is one of the ends of the chord through the pole of the plane containing the given three.

The vanishing of J requires the four contact points to be two harmonic pairs (see § 14).

For the discriminant we have

$$I^3 - 27J^2 = -\tfrac{1}{48}\Pi\Pi';$$

the vanishing of the discriminant makes two of the four tangents coincident; the condition for this is that one transversal meets the curve in a point P and the other transversal lies in the osculating plane at P.

The Hessian H of the quartic $(a, b, c, d, e \between \theta, 1)^4$ is given by

$$H \equiv \begin{vmatrix} a\theta^2 + 2b\theta + c, & b\theta^2 + 2c\theta + d \\ b\theta^2 + 2c\theta + d, & c\theta^2 + 2d\theta + e \end{vmatrix},$$

and for the quartic

$$f(\theta) \equiv n\theta^4 - 2m\theta^3 + (l+L)\,\theta^2 + \tfrac{2}{3}M\theta + \tfrac{1}{3}N,$$

we find

$$36H \equiv (A, B, C, D, E \between 1, -\theta)^4 - 3\,(l-L)\,f(\theta),$$

where A, B, C, D, E are the expressions of § 15. Now the pencil of quartics $f + \kappa H$, where κ is a parameter, have the following property: any set of four points given by a quartic of the pencil may be grouped in two pairs in three different ways; each grouping of two pairs determines an involution and the double points of the involution are given by one of the three quadratic factors of the sextic covariant of $f(\theta)$. This aspect will be further considered in §§ 22—25 and for the present we shall only remark that two members of the pencil of quartics have $f(\theta)$ as their Hessian. The expressions A, B, C, D, E satisfy the relations $AC - B^2 \equiv -3N\Pi$, etc. (§ 15), and these relations show that the Hessian of $(A, B, C, D, E \between 1, -\theta)^4$ is a multiple of $f(\theta)$. The other quartic of the pencil having $f(\theta)$ as its Hessian is

$$(A', B', C', D', E' \between 1, -\theta)^4,$$

where $A', B', \ldots$ are derived from $A, B, \ldots$ by interchanging l and L. This result gives another interpretation of the expressions

$$(A, B, C, D, E \between 1, -\theta)^4 \text{ and } (A', B', C', D', E' \between 1, -\theta)^4,$$

namely:

Given any line λ, there are four points whose osculating tetrahedral complexes of chords (§ 17) contain λ as a ray, these four points being given by $(A, B, C, D, E \between 1, -\theta)^4 = 0$; in the same way there are four points whose osculating tetrahedral complexes of lines-in-two-planes contain λ as a ray, and these four points are given by

$$(A', B', C', D', E' \between 1, -\theta)^4 = 0.$$

These two sets of four points have the same Hessian points, and the tangents at these Hessian points meet the given line λ.

Finally consider the case where the quartic $(a, b, c, d, e \between \theta, 1)^4$ is the square of a quadratic; in this case the quartic is a multiple of its Hessian, say $H = \kappa f(\theta)$. In order to find κ, we make use of the fact that the I-invariant of the Hessian is $\frac{1}{12}I^2$, so that $\frac{1}{12}I^2 = \kappa^2 I$, and therefore $6H + \sqrt{3I}.f \equiv 0$. Hence, on substituting the coefficients of the quartic, we have $12H \pm (l - L) f(\theta) \equiv 0$. Taking the upper sign, the conditions (equivalent to two only) are that A, B, C, D, E vanish, and in this case λ is a chord of the cubic; it meets the two pairs of coincident tangents at its ends. Taking the lower sign, A', B', C', D', E' all vanish and λ is a line-in-two-planes; it meets the two pairs of coincident tangents lying in the two osculating planes.

20. A screw associated with four points of a twisted cubic*.

Taking the parameters of the points as the roots of the quartic

$$n_1\theta^4 - 2m_1\theta^3 + (l_1 + L_1)\theta^2 + \tfrac{2}{3}M_1\theta + \tfrac{1}{3}N_1 = 0,$$

the two transversals of the four tangents have coordinates

$$(l_1, m_1, n_1, L_1, M_1, N_1), \quad (L_1, m_1, n_1, l_1, M_1, N_1).$$

The equations of two complexes containing the four tangents are

$$l - L = 0,$$

* Sturm, *Liniengeometrie.*

and $$\varpi_1 \equiv L_1 l + M_1 m + N_1 n + l_1 L + m_1 M + n_1 N = 0,$$
so that any screw containing the four tangents has for its equation
$$\varpi_1 + \phi(l - L) = 0,$$
where ϕ is a parameter.

Among these screws there is one which is reciprocal (apolar) to $l - L = 0$; the value of ϕ for this screw is given by
$$2\phi = l_1 - L_1.$$

Hence associated with the four points whose tangents meet λ_1 is the screw
$$K_1 \equiv 2\varpi_1 + (l_1 - L_1)(l - L) = 0.$$

Now take a second tetrad of points the tangents at which meet λ_2, the associated screw is
$$K_2 \equiv 2\varpi_2 + (l_2 - L_2)(l - L) = 0.$$

The screw K_1 contains the two transversals of the second tetrad of tangents if
$$2\varpi_{12} + (l_1 - L_1)(l_2 - L_2) = 0,$$
or, equivalently, $2(m_1 M_2 + m_2 M_1 + n_1 N_2 + n_2 N_1) + (l_1 + L_1)(l_2 + L_2) = 0$ which is the condition of apolarity of the two quartics giving the points of contact.

The result may be stated briefly thus:

Associated with four tangents to a cubic is a unique screw K_1 containing them and reciprocal to the screw containing all the tangents of the cubic: a set of four points apolar to these contact points has a similarly associated screw K_2. Then the two screws are reciprocal and each contains the two transversals of the four tangents belonging to the other screw.

The screw associated with the four points whose parameters are given by $(a, b, c, d, e \between \theta, 1)^4 = 0$ has for its equation
$$3cl + 6dm + 3en + 3cL - 2bM + aN = 0.$$

21. A system of tetrahedra whose vertices lie on one twisted cubic and whose faces osculate another twisted cubic.

Take the parametric equations of the twisted cubic to be
$$x : y : z : t = \theta^3 : \theta^2 : \theta : 1,$$
then any plane p cuts the cubic in three points whose parameters are given by
$$X\theta^3 + Y\theta^2 + Z\theta + T = 0.$$

Suppose these three points are three of the four points determined by the quartic equation

$$F(\theta) \equiv a\theta^4 + b\theta^3 + c\theta^2 + d\theta + e;$$

then if the remaining point has parameter ϕ, we have

$$X : Y : Z : T = a : a\phi + b : a\phi^2 + b\phi + c : a\phi^3 + b\phi^2 + c\phi + d,$$

and therefore, if ϕ is the parameter determining any one of the vertices of the tetrahedron given by $F(\theta) = 0$, the equation of the opposite face is

$$p \equiv a(x + \phi y + \phi^2 z + \phi^3 t) + b(y + \phi z + \phi^2 t) + c(z + \phi t) + dt = 0.$$

Now consider the system of ∞^1 tetrahedra whose vertices are given by the pencil of quartics

$$F_1(\theta) + \mu F_2(\theta) \equiv (a_1\theta^4 + b_1\theta^3 + c_1\theta^2 + d_1\theta + e_1) + \mu(a_2\theta^4 + b_2\theta^3 + c_2\theta^2 + d_2\theta + e_2) = 0,$$

where μ is a parameter. If ϕ is the parameter of a vertex of the tetrahedron given by $F_1 + \mu F_2 = 0$, the equation of the opposite face is $p_1 + \mu p_2 = 0$, where p_1 is derived from the expression for p above by writing $a_1, b_1, \ldots$ in place of $a, b, \ldots$. As μ varies the equation of the face of the variable tetrahedron is

$$\begin{vmatrix} p_1 & p_2 \\ F_1(\phi) & F_2(\phi) \end{vmatrix} = 0,$$

and this equation may be written in the form

$$\begin{vmatrix} a_1 & b_1 & c_1 & d_1 & e_1 \\ a_2 & b_2 & c_2 & d_2 & e_2 \end{vmatrix} \begin{vmatrix} x + \phi y + \phi^2 z + \phi^3 t & y + \phi z + \phi^2 t & z + \phi t & t & . \\ \phi^4 & \phi^3 & \phi^2 & \phi & 1 \end{vmatrix} = 0.$$

Writing $(ab) \equiv a_1 b_2 - a_2 b_1$, etc., the expanded form of this equation is

$$\begin{aligned} &\phi^3 \{(ab)x + (ac)y + (ad)z + (ae)t\} \\ &+ \phi^2 \{(ac)x + \overline{(ad) + (bc)}\,y + \overline{(ae) + (bd)}\,z + (be)t\} \\ &+ \phi\ \{(ad)x + \overline{(ae) + (bd)}\,y + \overline{(be) + (cd)}\,z + (ce)t\} \\ &+ \ \{(ae)x + (be)y + (ce)z + (de)t\} = 0. \end{aligned}$$

The coordinates of this plane are cubic functions of the parameter ϕ, and therefore the faces of the ∞^1 tetrahedra whose vertices lie on the given cubic osculate another twisted cubic. Moreover the face of the variable tetrahedron opposite the vertex with parameter ϕ is the polar plane of that vertex with regard to the fixed quadric whose equation is

$$\big((ab), (ad) + (bc), (be) + (cd), (de), (ae) + (bd), (ad), (ac), (ae), (be), (ce)\big)\!\big(x, y, z, t\big)^2 = 0,$$

so that each of the variable tetrahedra is self-polar with respect to this quadric, and the cubic locus of the vertex reciprocates into the cubic osculating the faces.

Any two tetrahedra of the system may be chosen arbitrarily: take the equations of two sets of vertices as $P_1=0$, $P_2=0$, $P_3=0$, $P_4=0$ and $P_1'=0$, $P_2'=0$, $P_3'=0$, $P_4'=0$. The equation of the quadric can be written in either of the forms $\sum_1^4 \kappa_1 P_1^2=0$, $\sum_1^4 \kappa_1' P_1'^2=0$, and therefore there is an identical relation $\sum_1^8 \kappa_1 P_1^2 \equiv 0$ connecting the equations of *any* eight points on a twisted cubic. This result and others associated with it have been discussed by P. Serret*.

The preceding properties are an extension of the theorem in plane geometry that, if sets of three points are given on a conic by the equation $a_\theta^3 + \mu a'_\theta{}^3 = 0$, where μ is a parameter, then the sides of the triangles so determined touch another conic, and the triangles are all self-conjugate with regard to a third conic.

22. Mutually inscribed tetrahedra associated with a twisted cubic.

Take any four points P_1, P_2, P_3, P_4 of a twisted cubic: p_1, p_2, p_3, p_4 are the osculating planes and λ_1, λ_2, λ_3, λ_4 are the tangents at these points; consider the four tetrahedra:

A. With vertices P_1, P_2, P_3, P_4.

B. With faces p_1, p_2, p_3, p_4.

C. With vertices at the points where λ_1, λ_2, λ_3, λ_4 meet the opposite faces of *A*.

D. With faces through λ_1, λ_2, λ_3, λ_4 and the corresponding vertices of *B*.

It will be shown that every pair of these tetrahedra are mutually inscribed; that is, taking any two such as *A* and *B*, the vertices of *A* lie on the faces of *B* and the vertices of *B* lie on the faces of *A*. This relation will be proved true for each of the six pairs of tetrahedra.

Take first the pair *A*, *B*; the vertices of *A* lie on the faces of *B* by hypothesis, and if the tetrahedron *A* is polarized with respect to the screw containing the tangents it becomes the tetrahedron *B*. It follows therefore that the vertices of *B* lie on the faces of *A*.

* *Géométrie de Direction*, Chap. IX

Denote by S_{AB} the operation of polarizing with respect to this screw, then, symbolically,

$$S_{AB}A = B, \quad S_{AB}B = A,$$

S_{AB} being a self-inverse operator, so that $S^2_{AB} \equiv 1$. The vertices and faces of either tetrahedron are polarized into the faces and vertices of the other.

The result will be established by finding screws which will polarize any one tetrahedron into any other, and, using the same notation, S_{CD} will represent the operation of polarizing the tetrahedron C into the tetrahedron D. Now these two tetrahedra are reciprocal with regard to the screw containing the tangents, so that

$$S_{AB} \equiv S_{CD},$$

and so

$$S_{AB}C = D, \quad S_{AB}D = C.$$

To find the remaining polarizing screws, take the four given points as the vertices of the tetrahedron of reference; then the parametric equations of the twisted cubic are (§ 2)

$$x : y : z : t = \frac{1}{\theta - a} : \frac{1}{\theta - b} : \frac{1}{\theta - c} : \frac{1}{\theta - d},$$

a, b, c, d being the values of the parameter θ at the four vertices.

The coordinates of the tangent at the point θ are found from the array

$$\left| \begin{array}{cccc} \dfrac{1}{\theta - a}, & \dfrac{1}{\theta - b}, & \dfrac{1}{\theta - c}, & \dfrac{1}{\theta - d} \\ \dfrac{1}{(\theta - a)^2}, & \dfrac{1}{(\theta - b)^2}, & \dfrac{1}{(\theta - c)^2}, & \dfrac{1}{(\theta - d)^2} \end{array} \right|,$$

and are given by

$$\frac{l}{(a-d)(\theta - b)^2(\theta - c)^2} = \frac{m}{(b-d)(\theta - c)^2(\theta - a)^2} = \frac{n}{(c-d)(\theta - a)^2(\theta - b)^2}$$
$$= \frac{L}{(b-c)(\theta - a)^2(\theta - d)^2} = \frac{M}{(c-a)(\theta - b)^2(\theta - d)^2} = \frac{N}{(a-b)(\theta - c)^2(\theta - d)^2}.$$

The equation of the screw containing the tangents is therefore

$$(a-d)^2(b-c)\,l + (b-d)^2(c-a)\,m + (c-d)^2(a-b)\,n$$
$$+ (b-c)^2(a-d)\,L + (c-a)^2(b-d)\,M + (a-b)^2(c-d)\,N = 0.$$

The operation of polarizing with regard to this screw has already been denoted by S_{AB}.

Consider the two tetrahedra A and C; the tangent at the point X, for which $\theta = a$, has coordinates given by

$$\frac{l}{(a-b)(a-c)} = \frac{m}{0} = \frac{n}{0} = \frac{L}{0} = \frac{M}{-(a-b)(a-d)} = \frac{N}{(a-c)(a-d)},$$

and it meets the plane $x=0$ in the point whose equation is

$$NY - MZ + lT = 0,$$

or

$$\frac{Y}{a-b} + \frac{Z}{a-c} + \frac{T}{a-d} = 0.$$

This is one of the vertices of the tetrahedron C, and it is the pole of $x=0$ with regard to the screw

$$\frac{l}{b-c} + \frac{m}{c-a} + \frac{n}{a-b} + \frac{L}{a-d} + \frac{M}{b-d} + \frac{N}{c-d} = 0.$$

The same screw gives the remaining vertices of C as the poles of the planes y, z, t of the tetrahedron A. The two tetrahedra are therefore mutually inscribed, and moreover the two screws are reciprocal, so that the order of polarizing with regard to them is commutative, and therefore

$$S_{AC} S_{AB} \equiv S_{AB} S_{AC}.$$

At this stage it is convenient to introduce a notation which expresses the equations in a simpler form.

We write

$$\alpha = (b-c)(a-d), \quad \beta = (c-a)(b-d), \quad \gamma = (a-b)(c-d),$$

so that $\alpha + \beta + \gamma \equiv 0$, and the cross-ratio of the vertices of the tetrahedron $A = (abcd) = -\beta/\alpha$; further we write

$$\lambda \equiv (a-d)l + (b-c)L, \quad \mu \equiv (b-d)m + (c-a)M, \quad \nu \equiv (c-d)n + (a-b)N,$$

so that $\lambda = 0$, $\mu = 0$, $\nu = 0$ are the equations of three mutually reciprocal screws. The polarizing screws already found can be written

$$S_{AB}: \quad \alpha\lambda + \beta\mu + \gamma\nu = 0,$$

$$S_{AC}: \quad \lambda/\alpha + \mu/\beta + \nu/\gamma = 0.$$

Next consider the two tetrahedra A and D; any face of D is to be found by operating with S_{AB} on a vertex of C, and therefore taking the vertex of C given by the equation

$$Y/(a-b) + Z/(a-c) + T/(a-d) = 0,$$

the corresponding face of D has for its equation

$$(a-b)\gamma(\alpha-\beta)y - (c-a)\beta(\gamma-\alpha)z + (a-d)\alpha(\beta-\gamma)t = 0.$$

This plane is the polar plane of X with regard to the screw

$$\alpha(\beta-\gamma)\lambda + \beta(\gamma-\alpha)\mu + \gamma(\alpha-\beta)\nu = 0,$$

and the same screw polarizes the other three faces of D into the vertices Y, Z, T of A The operation of polarizing with respect to this screw will therefore be denoted by S_{AD}; also the screw is reciprocal to each of the two already found and therefore the three operators S_{AB}, S_{AC}, S_{AD} are commutative. Hence

$$S_{AD}B = S_{AB}^2 S_{AD}B = S_{CD} S_{AD} S_{AB} B = C,$$

so that the tetrahedra B, C are mutually inscribed and $S_{BC} \equiv S_{AD}$.

Also $$S_{AC}B = S_{AB}^2 S_{AC}B = S_{CD} S_{AC} S_{AB} B = D,$$

so that the tetrahedra B, D are mutually inscribed and $S_{AC} \equiv S_{BD}$.

The mutual inscription of every pair has now been established: the screw which polarizes any two of the tetrahedra into each other also polarizes the remaining two into each other, and the three screws with this property are mutually reciprocal.

23. Screws related to the mutually inscribed tetrahedra.

The screw $\alpha(\beta-\gamma)\lambda + \beta(\gamma-\alpha)\mu + \gamma(\alpha-\beta)\nu = 0$ is reciprocal to the screw containing the tangents, and, since it polarizes the tetrahedra A and D of § 22 into each other, it has λ_1, λ_2, λ_3, λ_4 as rays; it is therefore the screw associated (in the sense of § 20) with the four vertices P_1, P_2, P_3, P_4 of the tetrahedron A.

It can be verified that any two chords of a twisted cubic and the four tangents at their ends are rays of a screw; for, taking the two chords as the edges yz, xt of the tetrahedron of reference, any screw containing the four tangents at their ends has equation

$$(\beta-\gamma+\kappa)\,\alpha\lambda + (\gamma-\alpha+\kappa)\,\beta\mu + (\alpha-\beta+\kappa)\,\gamma\nu = 0,$$

where κ is a parameter, and if $\beta-\gamma+\kappa=0$ the lines yz, xt are rays of the screw and its equation is $\mu-\nu=0$. The result bears some analogy to the theorem that two chords of a conic and the four tangents at their ends are tangents to another conic.

Consider the three screws given by the equations

$$\lambda' \equiv (a-d)\,l - (b-c)\,L = 0,$$

$$\mu' \equiv (b-d)\,m - (c-a)\,M = 0,$$

$$\nu' \equiv (c-d)\,n - (a-b)\,N = 0.$$

They are mutually reciprocal and also each of them is reciprocal to each of the three polarizing screws of § 22, so that the six screws are mutually reciprocal. The rays common to the three screws λ', μ', ν' are generators of a quadric, whose locus equation is found by eliminating $l : m : n$ from the equations

$$(a-d)\,lt-(b-c)\,(mz-ny)=0,$$
$$(b-d)\,mt-(c-a)\,(nx-lz)=0,$$
$$(c-d)\,nt-(a-b)\,(ly-mx)=0.$$

The equation of the quadric is

$$(a-b)\,(a-c)\,(a-d)\,x^2+\ldots=0,$$

or
$$f'(a)\,x^2+f'(b)\,y^2+f'(c)\,z^2+f'(d)\,t^2=0,$$

where
$$f(\theta)\equiv(\theta-a)\,(\theta-b)\,(\theta-c)\,(\theta-d).$$

The generators of the other system of this quadric are the rays common to the screws λ, μ, ν and common therefore to the three polarizing screws of § 22. Successive polarization with respect to these three polarizing screws, that is the operation denoted by $S_{AB}\,S_{AC}\,S_{AD}$, is equivalent to successive polarization with respect to the three screws λ', μ', ν', and is also equivalent to polarization with respect to this quadric, so that each of the tetrahedra A, B, C, D of § 22 is self-polar with respect to this quadric.

24. The Hessian and the sextic covariant of a binary quartic.

The sextic covariant of the quartic

$$f(\theta)\equiv(\theta-a)\,(\theta-b)\,(\theta-c)\,(\theta-d)$$

has three quadratic factors u, v, w, where

$$u\equiv(b-d)\,(\theta-c)\,(\theta-a)+(c-a)\,(\theta-b)\,(\theta-d)$$
$$\equiv(c-d)\,(\theta-a)\,(\theta-b)-(a-b)\,(\theta-c)\,(\theta-d),$$

and the corresponding expressions for v, w are derived by interchanging a, b, c cyclically. The two points given by $u=0$ are the double points of the involution determined by the two pairs of points $\theta=b$, $\theta=c$, and $\theta=a$, $\theta=d$; and further the three pairs of points given by $u=0$, $v=0$, $w=0$ are mutually harmonic. The quadratic factors u, v, w are related by the identity $\alpha u^2+\beta v^2+\gamma w^2\equiv 0$, where $\alpha\equiv(b-c)\,(a-d)$, etc., as in § 22. In terms of u, v, w the original quartic $f(\theta)$ is equivalent to any one of the forms v^2-w^2, w^2-u^2, u^2-v^2, and its Hessian is a multiple of $u^2+v^2+w^2$.

The coordinates of the chord joining the points of the cubic given by $u=0$ are found to be $\{-(b-c), (c-a), (a-b), -(a-d), (b-d), (c-d)\}$, and the equation of the chord can therefore be written in the form $-\lambda+\mu+\nu=0$, where λ, μ, ν are the expressions of § 22. The equations of the chords determined by $v=0$, $w=0$ are similarly

$$\lambda-\mu+\nu=0, \quad \lambda+\mu-\nu=0.$$

Associated with these three chords is the line whose equation is $\lambda+\mu+\nu=0$; it will be seen presently that this line is one of the two transversals of the four tangents at the Hessian points. The four lines $\lambda \pm \mu \pm \nu=0$ are generators of the quadric

$$f'(a)\,x^2+f'(b)\,y^2+f'(c)\,z^2+f'(d)\,t^2=0$$

determined (§ 23) by the three screws λ, μ, ν, and the geometrical relations between the four lines may be stated thus: a unique quadric can be described having as generators the line $\lambda+\mu+\nu=0$ and the two edges yz, xt of the tetrahedron of reference; another generator of the same system is $-\lambda+\mu+\nu=0$, and these four generators are cut harmonically by any generator of the other system. We may say that the lines $\pm\lambda+\mu+\nu=0$ are harmonic conjugates with respect to the two lines yz, xt. In the same way, the lines $\lambda \pm (\mu-\nu)=0$ are harmonic conjugates with respect to yz, xt; the lines $\lambda \pm \mu+\nu=0$ are harmonic conjugates with respect to zx, yt, and so on.

Any screw has four rays which are tangents to the cubic; the parameters of the points of contact are to be found by writing (§ 22)

$$l=(a-d)\,(\theta-b)^2\,(\theta-c)^2, \quad L=(b-c)\,(\theta-a)^2\,(\theta-d)^2, \quad \text{etc.}$$

in the equation of the screw. With these substitutions, using the relations

$$(a-d)\,(\theta-b)\,(\theta-c)=\tfrac{1}{2}\,(v+w), \ (b-c)\,(\theta-a)\,(\theta-d)=-\tfrac{1}{2}\,(v-w), \ \text{etc.,}$$

we have $\lambda=\tfrac{1}{2}\,(v^2+w^2), \ \mu=\tfrac{1}{2}\,(w^2+u^2), \ \nu=\tfrac{1}{2}\,(u^2+v^2),$

and therefore $-\lambda+\mu+\nu=u^2$, etc.

We shall now consider in succession the sets of four tangents which are rays of the screws related to the tetrahedra of § 22.

(i) The screw containing the tangents is given by

$$\alpha\lambda+\beta\mu+\gamma\nu=0$$

and this leads to the identity $\alpha u^2+\beta v^2+\gamma w^2 \equiv 0$.

(ii) The contact points of the four tangents contained in the screw

$$\alpha\,(\beta-\gamma)\,\lambda+\beta\,(\gamma-\alpha)\,\mu+\gamma\,(\alpha-\beta)\,\nu=0$$

are given by $\alpha(\beta-\gamma)u^2+\beta(\gamma-\alpha)v^2+\gamma(\alpha-\beta)w^2=0$, or equivalently $u^2=v^2=w^2$. This is the same as $f(\theta)=0$ and the four points are the vertices of the tetrahedron A.

(iii) The contact points of the four tangents contained in the screw $\lambda/\alpha+\mu/\beta+\nu/\gamma=0$ are given by $\alpha^2u^2+\beta^2v^2+\gamma^2w^2=0$, and this quartic is the unique member of the pencil of quartics, determined by $f(\theta)$ and its Hessian, which is apolar to $f(\theta)$.

(iv) The line $\lambda+\mu+\nu=0$ meets the tangents at the four points given by $u^2+v^2+w^2=0$, the Hessian of $f(\theta)$. The line $-\lambda+\mu+\nu=0$ meets the tangents at the points given by $u^2=0$, and this agrees with the fact that $-\lambda+\mu+\nu=0$ is the chord whose ends are given by $u=0$.

(v) The screw $\mu-\nu=0$ meets the tangents at the points given by $v^2-w^2=0$, or by $f(\theta)=0$, in agreement with § 23.

(vi) The screw $\lambda'\equiv(a-d)l-(b-c)L=0$ contains the four tangents whose contact points are given by

$$(a-d)(\theta-b)(\theta-c)\mp(b-c)(\theta-a)(\theta-d)=0,$$

that is by $vw=0$; the four points are the ends of the chords

$$\lambda\pm(\mu-\nu)=0.$$

This equation for θ may be written in the form

$$\{1/(\theta-a)-1/(\theta-d)\}\mp\{1/(\theta-b)-1/(\theta-c)\}=0,$$

so that the planes given by $x-t\mp(y-z)=0$ pass through the point $\theta=\infty$ of the cubic and the chords whose ends are given respectively by $v=0$, $w=0$. There are three such planes whose equations are

$$x-y-z+t=0,\quad -x+y-z+t=0,\quad -x-y+z+t=0.$$

Finally, since the five screws given by

$$\lambda'=0,\ \mu'=0,\ \nu'=0,$$

$$(\beta-\gamma)\alpha\lambda+(\gamma-\alpha)\beta\mu+(\alpha-\beta)\gamma\nu=0,\ \lambda/\alpha+\mu/\beta+\nu/\gamma=0$$

are mutually reciprocal, the four quartics

$$vw,\ wu,\ uv,\ \alpha^2u^2+\beta^2v^2+\gamma^2w^2$$

and the original quartic $f(\theta)$, which is equivalent to any one of the forms v^2-w^2, w^2-u^2, u^2-v^2, constitute a system of five mutually apolar quartics.

25. A special system of tetrahedra inscribed in the cubic.

In § 21 we considered a system of tetrahedra determined by a pencil of quartics. The pencil of quartics may be of the special form $f+\kappa H$, where $f(\theta)\equiv(\theta-a)(\theta-b)(\theta-c)(\theta-d)$ and H is the Hessian of f. Any tetrahedron of this system has each pair of opposite edges as chords whose ends belong to one of three involutions whose double points are given by $u=0$, or $v=0$, or $w=0$. Consider the involution whose double points are given by $u=0$; the chords are all generators of a quadric (§ 10) and among these generators are the edges yz, xt of the tetrahedron of reference and the lines $\lambda\pm(\mu-\nu)=0$, which (§ 23) join the pairs of points given by $v=0$, $w=0$. Any other generator of the system has equation

$$(b-c)\,L/\phi+\phi\,(a-d)\,l+\mu-\nu=0,$$

where ϕ is a parameter, since the values 0, ∞, ± 1 of ϕ give the four generators already mentioned. This generator is an edge of the variable tetrahedron, and the equation of the opposite edge is

$$\phi\,(b-c)\,L+(a-d)\,l/\phi+\mu-\nu=0.$$

Subtraction of the equations of the two generators shows them to be polar lines with respect to the screw

$$\lambda'\equiv(a-d)\,l-(b-c)\,L=0.$$

The equation of the quadric generated by the chords of the involution is

$$(by-cz)\,(x-t)-(ax-dt)\,(y-z)=0.$$

The other two pairs of edges of the variable tetrahedron corresponding to involutions with double points given by $v=0$, or $w=0$, are given by similar equations.

It was proved in § 21 that the variable tetrahedra are all self-polar with respect to a certain quadric: in order to determine this quadric in the present case we may assume for its equation

$$\xi x^2+\eta y^2+\zeta z^2+\tau t^2=0,$$

since the tetrahedron of reference is self-polar with respect to it; the coefficients ξ, η, ζ, τ are to be found. If θ_1, θ_2 are the parameters of two points given by the Hessian, we have, since these points are conjugate with respect to the quadric,

$$\xi/(\theta_1-a)\,(\theta_2-a)+\eta/(\theta_1-b)\,(\theta_2-b)$$
$$+\zeta/(\theta_1-c)\,(\theta_2-c)+\tau/(\theta_1-d)\,(\theta_2-d)=0,$$

and therefore

$$\xi/(\theta_1-a)+\eta/(\theta_1-b)+\zeta/(\theta_1-c)+\tau/(\theta_1-d)$$
$$=\xi/(\theta_2-a)+\eta/(\theta_2-b)+\zeta/(\theta_2-c)+\tau/(\theta_2-d).$$

Hence the values of the expression

$$\xi/(\theta-a)+\eta/(\theta-b)+\zeta/(\theta-c)+\tau/(\theta-d)\equiv\kappa, \text{ say,}$$

are the same when the four roots of the Hessian are substituted successively for θ. The Hessian of $f(\theta)$ is a multiple of

$$H(\theta)\equiv\Sigma\,(a-d)^2(\theta-b)^2(\theta-c)^2,$$

and, therefore, a multiplier ρ, independent of θ, can be found such that

$$\xi(\theta-b)(\theta-c)(\theta-d)+\ldots\ldots\equiv\rho H(\theta)+\kappa f(\theta).$$

In this identity, put $\theta=a$, then

$$\xi=3\rho\,(a-b)(a-c)(a-d),$$

and the values of η, ζ, τ are found similarly by putting $\theta=b$, c, d in turn.

The equation of the polarizing quadric is therefore

$$f'(a)\,x^2+f'(b)\,y^2+f'(c)\,z^2+f'(d)\,t^2=0\,;$$

this quadric has already appeared (§ 23) as the quadric with regard to which each of the tetrahedra A, B, C, D of § 22 is self-polar.

Let s ($\equiv S_{AB}\,S_{AC}\,S_{AD}$ in the notation of § 22) denote the operation of polarizing with regard to this quadric, and let A' denote any inscribed tetrahedron of the system we have been considering; then

$$sA\equiv A,\ sB\equiv B,\ sC\equiv C,\ sD\equiv D,\qquad(§\ 22)$$

and we have also proved that $sA'\equiv A'$. Associated with the tetrahedron A' there are three tetrahedra B', C', D' derived from it in the same way that B, C, D are derived from A, and it will now be proved that these tetrahedra B', C', D' are also self-polar with regard to the same quadric. The method of deriving B' from A' shows that $B'\equiv S_{AB}A'$, and therefore

$$sB'\equiv sS_{AB}A'\equiv S_{AB}\,sA'\equiv S_{AB}A'\equiv B',$$

so that the tetrahedron B' is self-polar with respect to the quadric. Since A' and B' are self-polar with respect to the quadric, the quadric is determined by them and therefore the remaining tetrahedra C' and D' are, as in § 23, self-polar with respect to the same quadric, so that $sC'\equiv C'$ and $sD'\equiv D'$.

Hence the special pencil of quartics $f+\kappa H$ determines four systems of tetrahedra of the nature prescribed in § 22 and all the tetrahedra are self-polar with regard to the same quadric.

26. Some geometrical definitions of the twisted cubic.

We shall now consider some of the conventional definitions of the twisted cubic and show how they lead to the analytical definition of § 1.

It will appear that in some cases we are in effect only proving the converses of theorems already established.

1. A twisted cubic is the residual intersection of two quadrics with a common generator.

(i) Consider first the case in which both the quadrics are cones; take two of the vertices X, T of the tetrahedron of reference at the vertices of the cones and take Y, Z as any two points on their curve of intersection. Since the two cones have XY, XZ, XT and TX, TY, TZ respectively as generators, their equations can be written in the forms

$$fyz + vyt + wzt = 0,$$

and

$$f'yz + g'zx + h'xy = 0.$$

To find the common points of the two cones (other than those lying on the common generator), take $y/z = \theta$ as a parameter, and then the equations

$$x : y : z : t = \frac{-f'\theta}{g' + h'\theta} : \theta : 1 : \frac{-f\theta}{w + v\theta}$$

show the residual intersection as a twisted cubic, unless $w/v = g'/h'$, in which case the cones touch along their common generator and have a conic as their residual intersection. If $\theta = -g'/h'$, the corresponding point of the curve is X, and the value $-w/v$ of the parameter gives the point T, so that the common generator is a chord of the cubic and the vertices of the cones are the ends of the chord.

The unique twisted cubic through six points P_1, P_2, P_3, P_4, P_5, P_6 is determined synthetically by taking two quadric cones one with vertex P_1 and containing the points $P_2, \ldots, P_6$, the other with vertex P_6 and containing the points $P_1, \ldots, P_5$. P_1P_6 is a common generator of these two cones and their residual intersection is the unique cubic through the six points. The six cones so determined all contain the cubic.

Specially the unique cubic touching three given lines at three given points is determined geometrically as follows: if λ_1, λ_2, λ_3 are the tangents and P_1, P_2, P_3 are their points of contact, there is a unique quadric cone which has P_1 as vertex, has λ_1 as generator and touches the planes $P_1\lambda_2$ and $P_1\lambda_3$ along P_1P_2 and P_1P_3 respectively. This cone contains the twisted cubic, which is therefore given as the common curve of this cone and the two corresponding cones with their vertices at P_2, P_3.

(ii) Next consider the case where the quadrics are not both cones; taking yz as the common generator, the equations of the quadrics can be written in the forms

$$xy - zt = 0,$$

and $$by^2 + cz^2 + 2fyz + 2gzx + 2hxy + 2vyt = 0,$$

by taking a tetrahedron of reference with two pairs of opposite edges as generators of the non-degenerate quadric. Any point on the first quadric has coordinates $(\theta, \phi, \theta\phi, 1)$, where θ and ϕ are parameters: this point lies on the second quadric if

$$b\phi^2 + c\theta^2\phi^2 + 2f\theta\phi^2 + 2g\theta^2\phi + 2h\theta\phi + 2v\phi = 0.$$

Rejecting the value $\phi = 0$, which gives points on the common generator yz, we have

$$-\phi = 2(g\theta^2 + h\theta + v)/(c\theta^2 + 2f\theta + b).$$

Hence the coordinates of any point on the residual intersection are given by

$$x : y : z : t = \theta(c\theta^2 + 2f\theta + b) : -2(g\theta^2 + h\theta + v) :$$
$$-2\theta(g\theta^2 + h\theta + v) : (c\theta^2 + 2f\theta + b),$$

in agreement with § 1. The chord of the cubic which joins the two points whose parameters are given by

$$g\theta^2 + h\theta + v = 0$$

is the common generator of the two quadrics.

Correlatively we can define a twisted cubic as the curve osculated by the common tangent planes of two quadrics with a common generator. Specially the two quadrics may degenerate into two conics touching at two distinct points the line of intersection of their planes. In this last case the tetrahedron of reference may be suitably chosen thus: take the points of contact of the two conics with the line as the vertices Y, Z; through Y draw the other tangent to the second conic and take the point of contact as X; and through Z draw the other tangent to the first conic and take the point of contact as T. Then the equations of the two conics can be written $ZX - Y^2 = 0$, $YT - Z^2 = 0$, and the coordinates of any common tangent plane are given by $X : Y : Z : T = \theta^3 : \theta^2 : \theta : 1$, where θ is a parameter. The correlative choice of a tetrahedron of reference in (i) leads immediately to the parametric locus equations $x : y : z : t = \theta^3 : \theta^2 : \theta : 1$.

2. A twisted cubic is the locus of points of intersection of corresponding planes in three homographic pencils of planes. (A pencil of planes is a single infinity of planes through a common line.)

Take the equations of the three pencils to be

$$p + kq = 0, \quad p' + k'q' = 0, \quad p'' + k''q'' = 0,$$

where k, k', k'' are parameters of which every two are related homographically. The coordinates of the common point are found by solving the three linear equations for $x : y : z : t$; these coordinates are proportional to cubic functions of any one of the parameters k, k', k''.

The present definition may be reduced to (1) by the following argument: corresponding planes of the pencils through pq, $p'q'$ intersect in lines generating a quadric, which has pq, $p'q'$ as generators. The pencils through pq, $p''q''$ generate similarly a quadric having pq, $p''q''$ as generators. The two quadrics have pq as a common generator and the required locus is their residual intersection. The axes of the three pencils are chords of the curve.

It follows from this definition that the locus of poles of a plane with respect to a system of quadrics $s + ks' = 0$ through a common quartic is a twisted cubic.

The correlative definition is "a twisted cubic is the curve osculated by the planes through corresponding points of any three homographic ranges," and, as a special case of this, the polar planes of a point with respect to a system of quadrics inscribed in a common developable osculate a twisted cubic.

3. A twisted cubic is the locus of the points of intersection of corresponding lines in two homographic bundles of lines. (A bundle of lines is a set of ∞^2 lines through a given point.)

In general two corresponding lines will not meet, but if the vertices of the bundles are P, P', it is known that the lines through P which meet their corresponding lines generate a quadric cone which contains P'; while the lines through P' which meet their corresponding lines generate a quadric cone which contains P. The required locus is therefore the residual intersection of two quadric cones with a common generator; the common line of the two bundles is a chord of the curve.

Correlatively we may define the twisted cubic as the curve osculated by planes through the corresponding (coplanar) rays of two homographic plane fields. (A plane field of rays is a set of ∞^2 coplanar lines.)

4. Among other projective modes of generating a twisted cubic may be mentioned the following:

(i) In two collineated spaces if P and P' are corresponding points collinear with a fixed point O, then the locus of P is a twisted cubic.

More specially: if two variable points P and P', collinear with a fixed point O, have the same polar plane with respect to two quadrics s and s' respectively, then P generates a twisted cubic.

And still more specially: the locus of a point P, such that the normal from P to its polar plane with regard to a quadric s passes through a fixed point O, is a twisted cubic, which passes through the feet of the six normals from O to the quadric.

(ii) Corresponding planes in three homographic sets of planes, of which one set form a pencil and each of the others tangent planes to a quadric cone, meet in a point generating a twisted cubic.

(iii) The intersections of corresponding planes in three homographic sets of osculating planes of a twisted cubic lie on another twisted cubic.

(iv) In a (1, 1) point correspondence in space any point P has two correspondents; if the plane through P and its two correspondents passes through a fixed line, then P generates a twisted cubic.

Various kinematical methods of generation, which may be included under these four cases, are given by Chasles in the *Aperçu Historique.*

II

THE CUBICAL HYPERBOLA

27. Classification of twisted cubics with regard to the plane at infinity.

The plane at infinity meets the curve in three points, and from the nature of these three points is derived* a metrical classification of twisted cubics. It has already been pointed out that twisted cubics are not distinguishable from one another by their projective properties.

The three infinity points of the curve may be

(1) all real and distinct: the curve is then called a *cubical hyperbola.*

* Cremona, in *Crelle* LVIII. The classification is ascribed to Seydewitz.

(2) one real and two imaginary: the curve is then called a *cubical ellipse.*

(3) one real and two coincident: the curve is then called a *cubical hyperbolic parabola.*

(4) all coincident: the curve is then called a *cubical parabola.*

We shall now consider the first of these curves—the cubical hyperbola—in some detail. Many of the results present a notable analogy to the properties of the plane hyperbola.

28. Asymptotes and circumscribing cylinders.

The asymptotes of the curve are the tangents at the three real infinity points. In § 26, 1, (i) it was shown that a twisted cubic is determined by three of its points and the tangents thereat, so that a knowledge of the three asymptotes of a cubical hyperbola is sufficient to determine it.

Using oblique Cartesian coordinates, the equations of any three lines can be written in the forms

$$\left.\begin{matrix} y+b=0 \\ z-c=0 \end{matrix}\right\}, \quad \left.\begin{matrix} z+c=0 \\ x-a=0 \end{matrix}\right\}, \quad \left.\begin{matrix} x+a=0 \\ y-b=0 \end{matrix}\right\}.$$

The three lines lie along the edges of a skew box (parallelepiped), whose faces are each parallel to two of the three lines; the origin is the centre of the box, the axes are parallel to its edges and the lengths of the three edges are $2a$, $2b$, $2c$. Let us modify the coordinates by taking a, b, c to be the units of measurement along the coordinate axes: the equations of the three asymptotes can then be written in the forms

$$\left.\begin{matrix} y+1=0 \\ z-1=0 \end{matrix}\right\}, \quad \left.\begin{matrix} z+1=0 \\ x-1=0 \end{matrix}\right\}, \quad \left.\begin{matrix} x+1=0 \\ y-1=0 \end{matrix}\right\}.$$

The scale of measurement along any other line depends only on the direction of the line.

A return to oblique Cartesian coordinates can be made at any stage of what follows by writing x/a, y/b, z/c for x, y, z.

The twisted cubic is projected from any point of itself by a quadric cone: the cone becomes a cylinder when its vertex is one of the three infinity points. Denoting the asymptotes by α_1, α_2, α_3, the cylinder with its generators parallel to α_1 is uniquely determined by the conditions that α_1 is a generator and that the cylinder touches α_2, α_3 at their infinity points, so that the planes through α_2, α_3 parallel to α_1 are the asymptotic planes of the cylinder.

The equation of the cylinder is therefore of the form

$$(y-1)(z+1)+k=0,$$

where k is to be determined from the remaining condition that a_1 is a generator. This requires $k=4$, and therefore the three circumscribing cylinders of the cubical hyperbola are

$$yz+y-z+3=0, \text{ with axis } y-1=0,\ z+1=0;$$
$$zx+z-x+3=0, \text{ with axis } z-1=0,\ x+1=0;$$
and
$$xy+x-y+3=0, \text{ with axis } x-1=0,\ y+1=0.$$

Each pair of cylinders have as their common intersection the twisted cubic and a common generator which is the infinity line in one of the planes x, y, z. The cylinder axes are the reflexions of the asymptotes in the origin, and lie along three other edges of the skew box.

29. The asymptotic planes.

The asymptotic planes of the cubical hyperbola are the osculating planes at its infinity points. It follows from § 5 that the asymptotic plane through a_1 touches the cylinder $yz+y-z+3=0$ along a_1: its equation is therefore $y-z+2=0$. The three asymptotic planes are given by the equations

$$y-z+2=0, \quad z-x+2=0, \quad x-y+2=0.$$

The three planes are parallel to the line $x=y=z$; they meet at the infinity point of this line in agreement with the result (§ 7) that the osculating planes at three points meet in a point coplanar with the three points.

The three asymptotic planes form a triangular prism with its edges parallel to the line $x=y=z$: it will be found convenient to associate with this triangular prism the various lines which are related to the cubical hyperbola. So far six edges of the skew box are accounted for. Three of them are asymptotes, three of them are axes of the circumscribing cylinders; the six edges form a skew hexagon with its opposite sides parallel and the sides are alternately asymptote and cylinder axis. The line $x=y=z$ is that diagonal of the box which meets none of these six edges and the asymptotic planes are the planes through the asymptotes parallel to this diagonal. The remaining six edges of the box, three through each of the points $(1, 1, 1)$, $(-1, -1, -1)$, will be associated with the curve in due course (§ 45, (i)).

30. Topography of the cubical hyperbola: (i) the asymptotes, the asymptotic planes and the cylinder axes.

Denoting the asymptotes by $\alpha_1, \alpha_2, \alpha_3$, the edges of the triangular prism of asymptotic planes by $\beta_1, \beta_2, \beta_3$ and the cylinder axes by $\gamma_1, \gamma_2, \gamma_3$, we have the equations:

$$\alpha_1 \equiv \begin{cases} y+1=0, \\ z-1=0. \end{cases} \quad \alpha_2 \equiv \begin{cases} z+1=0, \\ x-1=0. \end{cases} \quad \alpha_3 \equiv \begin{cases} x+1=0, \\ y-1=0. \end{cases}$$

$$\beta_1 \equiv \begin{cases} z-x+2=0, \\ x-y+2=0. \end{cases} \quad \beta_2 \equiv \begin{cases} x-y+2=0, \\ y-z+2=0. \end{cases} \quad \beta_3 \equiv \begin{cases} y-z+2=0, \\ z-x+2=0. \end{cases}$$

$$\gamma_1 \equiv \begin{cases} y-1=0, \\ z+1=0. \end{cases} \quad \gamma_2 \equiv \begin{cases} z-1=0, \\ x+1=0. \end{cases} \quad \gamma_3 \equiv \begin{cases} x-1=0, \\ y+1=0. \end{cases}$$

The coordinates of the intersections of the asymptotes with the edges of the prism are given by

	β_1	β_2	β_3
α_1		$(-3, -1, 1)$	$(3, -1, 1)$
α_2	$(1, 3, -1)$		$(1, -3, -1)$
α_3	$(-1, 1, -3)$	$(-1, 1, 3)$	

The plane through the three points $\alpha_2\beta_3$, $\alpha_3\beta_1$, $\alpha_1\beta_2$ is given by

$$x+y+z+3=0;$$

and the plane through the three points $\beta_2\alpha_3$, $\beta_3\alpha_1$, $\beta_1\alpha_2$ is given by

$$x+y+z-3=0.$$

These planes are parallel and therefore make equal intercepts on the three lines $\beta_1, \beta_2, \beta_3$: hence any edge of the prism is cut by two asymptotes in two points whose distance apart is the same for all three edges of the prism. If the curve is placed so that the plane $x+y+z=0$ is horizontal, then as we go cyclically round the prism from face to face the asymptotes all slope upwards or all slope downwards.

From what precedes it follows that, if on the edges of any triangular prism we take three equal lengths AA', BB', CC' measured in the same sense, then there is a unique cubical hyperbola having BC', CA', AB' as its asymptotes, and the faces of the prism are the asymptotic planes.

Taking any triangular prism and two arbitrary lines one in each of two faces, we can construct a cubical hyperbola with these lines as two asymptotes and the faces of the prism as asymptotic planes.

On the asymptote a_1 are four notable points, namely

$$a_1\beta_2, \qquad a_1\gamma_2,$$
$$a_1\gamma_3, \qquad a_1\beta_3,$$

with coordinates

$$(-3, -1, 1), \quad (-1, -1, 1),$$
$$(1, -1, 1), \quad (3, -1, 1);$$

and therefore the intercept made on a_1 by the edges of the prism is trisected by the cylinder axes γ_2, γ_3.

The plane $x + y + z = 0$ bisects

(1) the intercept made by the prism edges on any asymptote,

(2) the intercept made on any prism edge by the two asymptotes, which meet it.

These relations are shown in figures 3, 4, where the six edges of the skew box are darkened: other associated lines will appear presently.

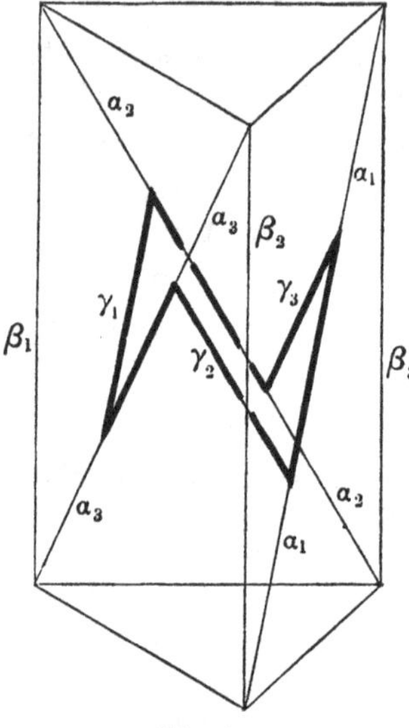

Fig. 3.

A model of the lines shown in the first figure is easily made by constructing a wire framework in the form of the triangular prism, taking

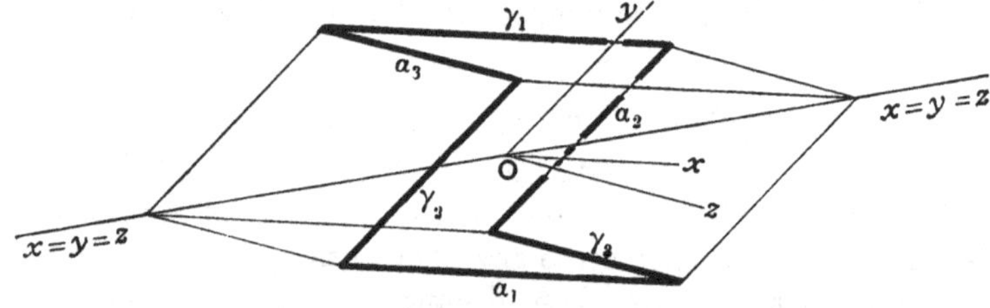

Fig. 4.

equal (arbitrary) lengths along its edges and introducing wires to represent the various lines. For the better understanding of what follows the reader is strongly recommended to construct a model* of this nature.

31. Parametric representation.

The twisted cubic is the common curve of the three circumscribing cylinders given by

$$\begin{cases} yz + y - z + 3 = 0, \\ zx + z - x + 3 = 0, \\ xy + x - y + 3 = 0, \end{cases}$$

* Some dimensions are given in § 58.

and so, for any point on the curve,

$$y = \frac{x+3}{1-x}, \quad z = \frac{x-3}{1+x};$$

writing $x = 1 - 2\theta$, we have

$$y = \frac{2-\theta}{\theta}, \quad z = \frac{\theta+1}{\theta-1}.$$

Hence parametric equations of the curve are

$$x = 1 - 2\theta, \quad y = \frac{2-\theta}{\theta}, \quad z = \frac{\theta+1}{\theta-1}.$$

This choice of the parameter ascribes to the infinity points the values ∞, 0, 1 of θ; θ may of course be replaced by $(a'\theta + b')/(c'\theta + d')$, where a', b', c', d' are arbitrary.

The equations of the three cylinders being cyclically symmetrical, it is natural to seek for parametric equations which shall also be cyclically symmetrical: this is effected by taking three quantities* α, β, γ such that

$$\alpha + \beta + \gamma = 0, \quad \theta = -\beta/\alpha.$$

Then

$$1 - 2\theta = \frac{\alpha + 2\beta}{\alpha} = \frac{\beta - \gamma}{\alpha},$$

$$\frac{2}{\theta} - 1 = -\frac{2\alpha + \beta}{\beta} = \frac{\gamma - \alpha}{\beta},$$

$$\frac{\theta + 1}{\theta - 1} = \frac{\beta - \alpha}{\alpha + \beta} = \frac{\alpha - \beta}{\gamma}.$$

So any point on the cubic is given by the symmetrical equations

$$x = \frac{\beta - \gamma}{\alpha}, \quad y = \frac{\gamma - \alpha}{\beta}, \quad z = \frac{\alpha - \beta}{\gamma},$$

where $\alpha + \beta + \gamma = 0$. Since the coordinates involve only the ratios $\alpha : \beta : \gamma$, there is only one effective parameter. The point on the curve with parameters (α, β, γ) and with coordinates

$$\left(\frac{\beta - \gamma}{\alpha}, \frac{\gamma - \alpha}{\beta}, \frac{\alpha - \beta}{\gamma}\right)$$

may without ambiguity be called "the point (α, β, γ)."

The three infinity points are given by $\alpha = 0$, $\beta = 0$, $\gamma = 0$.

* It is hardly necessary to point out that there is no connection between these parameters and the *names* of the lines associated with the curve.

If P is any point on the curve, then

$$x+1=\frac{\beta-\gamma}{a}+1=-\frac{2\gamma}{a}, \text{ etc.},$$

$$x-1=\frac{\beta-\gamma}{a}-1=\frac{2\beta}{a}, \text{ etc.},$$

and therefore
$$(x+1)(y+1)(z+1)+8=0,$$
$$(x-1)(y-1)(z-1)-8=0,$$

whence
$$xyz+x+y+z=0,$$
$$yz+zx+xy+9=0.$$

These are the equations of a cubic surface and a quadric circumscribing the cubical hyperbola: their relations to the curve will appear later (§§ 40, 43, 49).

32. Symmetrical parameters.

The cyclic symmetry of the curve suggested the use of the parameters a, β, γ; any expression involving these parameters admits of an infinite number of equivalent forms by virtue of the relation $a+\beta+\gamma=0$, and in a few instances it may be better to use the single unsymmetrical parameter θ. Passage from a, β, γ to θ, and reversely, is effected by the equations

$$a+\beta+\gamma=0, \quad \theta=-\beta/a.$$

An involution of points on the curve is given by an equation of the form

$$A'\theta\theta'+B'(\theta+\theta')+C'=0;$$

expressed in terms of parameters (a, β, γ), (a', β', γ') this equation is

$$A'\beta\beta'-B'(a\beta'+a'\beta)+C'aa'=0.$$

Also

$$(a\beta'+a'\beta)=(a+\beta)(a'+\beta')-aa'-\beta\beta'=\gamma\gamma'-aa'-\beta\beta',$$

and therefore the equation of the involution is

$$(B'+C')aa'+(A'+B')\beta\beta'-B'\gamma\gamma'=0,$$

or
$$Aaa'+B\beta\beta'+C\gamma\gamma'=0,$$

where the ratios $A:B:C$ are arbitrary.

Conversely any equation of this form represents an involution of points on the cubical hyperbola.

The double points of the involution are determined from the equations $Aa^2+B\beta^2+C\gamma^2=0$, $a+\beta+\gamma=0$.

The cross-ratio of four points in the sense determined by § 6 is equal to the cross-ratio of their four parameters; specially the cross-ratio of the three infinity points with parameters ∞, 0, 1 and the point with parameter θ is

$$\{\infty, 0, 1, \theta\} = \theta;$$

the harmonic conjugates of each infinity point in turn with respect to the other two are given by the values $\frac{1}{2}$, 2, -1 of the parameter θ.

The cross-ratio of four points on the curve with parameters $(\alpha_1, \beta_1, \gamma_1)$, $(\alpha_2, \beta_2, \gamma_2)$, $(\alpha_3, \beta_3, \gamma_3)$, $(\alpha_4, \beta_4, \gamma_4)$ is equal to

$$(\alpha_1\beta_3 - \alpha_3\beta_1)(\alpha_2\beta_4 - \alpha_4\beta_2)/(\alpha_1\beta_4 - \alpha_4\beta_1)(\alpha_2\beta_3 - \alpha_3\beta_2),$$

and this expression is not symmetrical. From the equations

$$\alpha_1 + \beta_1 + \gamma_1 = 0,$$
$$\alpha_3 + \beta_3 + \gamma_3 = 0,$$

we find

$$\beta_1\gamma_3 - \beta_3\gamma_1 = \gamma_1\alpha_3 - \gamma_3\alpha_1 = \alpha_1\beta_3 - \alpha_3\beta_1 = \mu_{13} = -\mu_{31}, \text{ say,}$$

and therefore the cross-ratio can be expressed symmetrically in the form

$$\mu_{13}\mu_{24}/\mu_{14}\mu_{23}.$$

33. The tangent at any point.

Since
$$x = \frac{\beta - \gamma}{\alpha},$$

$$dx = -\frac{\beta-\gamma}{\alpha^2}d\alpha + \frac{d\beta - d\gamma}{\alpha}$$
$$= \frac{(\alpha d\beta - \beta d\alpha) + (\gamma d\alpha - \alpha d\gamma)}{\alpha^2}$$
$$= \frac{2\mu}{\alpha^2},$$

where
$$\mu = \beta d\gamma - \gamma d\beta = \gamma d\alpha - \alpha d\gamma = \alpha d\beta - \beta d\alpha.$$

So
$$dx : dy : dz = \beta^2\gamma^2 : \gamma^2\alpha^2 : \alpha^2\beta^2,$$

and the coordinates of the tangent at the point (α, β, γ), found from the array

$$\begin{vmatrix} \beta^2\gamma^2, & \gamma^2\alpha^2, & \alpha^2\beta^2, & 0 \\ \frac{\beta-\gamma}{\alpha}, & \frac{\gamma-\alpha}{\beta}, & \frac{\alpha-\beta}{\gamma}, & 1 \end{vmatrix},$$

are given by

$$\frac{l}{\beta^2\gamma^2} = \frac{m}{\gamma^2\alpha^2} = \frac{n}{\alpha^2\beta^2} = \frac{L}{-\alpha^2(\alpha^2 + 2\beta\gamma)} = \frac{M}{-\beta^2(\beta^2 + 2\gamma\alpha)} = \frac{N}{-\gamma^2(\gamma^2 + 2\alpha\beta)}.$$

Hence

$$(L + M + N)/(l + m + n) = -(\alpha^4 + \beta^4 + \gamma^4)/(\beta^2\gamma^2 + \gamma^2\alpha^2 + \alpha^2\beta^2) = -2,$$

and the equation of the unique screw (§ 8) containing the tangents is

$$2(l + m + n) + L + M + N = 0.$$

34. The vertices, centre, central plane, and axis of a cubical hyperbola.

A *vertex* is defined as a finite point of the curve at which the tangent is parallel to an asymptotic plane.

If the tangent at (α, β, γ) is parallel to the plane $y - z + 2 = 0$, then

$$\gamma^2 \alpha^2 = \alpha^2 \beta^2,$$

and therefore, either

(i) $\alpha = 0$, or equivalently $\beta + \gamma = 0$, giving an infinity point, the tangent at which lies in the asymptotic plane; or

(ii) $\beta - \gamma = 0$, giving the point with coordinates $(0, 3, -3)$.

The value of the parameter $\theta = -\beta/\alpha$ for this point is $\frac{1}{2}$, and therefore this vertex is the harmonic conjugate of the infinity point on the asymptote a_1 with regard to the other two infinity points.

Hence the vertices of the curve may be defined alternatively as the three points of the curve with the property that each is the harmonic conjugate of one infinity point with regard to the other two infinity points. The previous definition may be derived from this by using the result of § 14; for, if A_1, A_2, A_3 are the infinity points and V_1 the harmonic conjugate of A_1 with respect to A_2, A_3, the tangent at V_1 meets the osculating plane at A_1 in a point on the plane $A_1A_2A_3$ which in the present instance is the plane infinity.

The three vertices of the cubical hyperbola have coordinates $(0, 3, -3)$, $(-3, 0, 3)$, $(3, -3, 0)$.

The *central plane* is the plane through the vertices: its equation is

$$x + y + z = 0.$$

The *centre* is the pole of the central plane with respect to the screw containing the tangents; its coordinates are $(0, 0, 0)$. The osculating planes at the vertices pass through it (§ 9).

The *axis* of the curve is the line through the centre parallel to the asymptotic planes; its equations are $x = y = z$ (see fig. 4). We shall see later (§ 44, (iii)) that the axis is also the unique chord of the cubic through the centre.

The plane through the axis and the edge β_1 of the triangular prism is given by $2x - y - z = 0$; it meets the asymptotic plane $y - z + 2 = 0$ in the line whose equations are

$$z - x - 1 = 0, \quad x - y - 1 = 0.$$

This line lies midway between the edges β_2, β_3; so the axis may be defined by the following property: any plane cuts the edges of the triangular prism formed by the asymptotic planes in three points whose centroid lies on the axis. The centroid of the three points $\alpha_2\beta_3$, $\alpha_3\beta_1$, $\alpha_1\beta_2$ has coordinates $(-1, -1, -1)$ and therefore in fig. 3 the centroids of the two triangular faces of the prism are the two remaining corners of the skew box not lying on the asymptotes.

35. The osculating plane at any point.

The polar plane of any point P_1 with respect to the screw containing the tangents is given by

$$(y_1 - z_1 + 2)\,x + (z_1 - x_1 + 2)\,y + (x_1 - y_1 + 2)\,z - 2\,(x_1 + y_1 + z_1) = 0.$$

If P_1 is the point (α, β, γ) of the curve, the polar plane is the osculating plane at P_1, and its equation reduces to

$$\alpha^3 x + \beta^3 y + \gamma^3 z + (\beta - \gamma)(\gamma - \alpha)(\alpha - \beta) = 0.$$

Hence the coordinates of any osculating plane are given parametrically by the equations

$$X : Y : Z : 1 = \alpha^3 : \beta^3 : \gamma^3 : (\beta - \gamma)(\gamma - \alpha)(\alpha - \beta),$$

where $\alpha + \beta + \gamma = 0$.

If $\alpha = 0$, $\beta = -\gamma$, the equation reduces to $y - z + 2 = 0$, the equation of the asymptotic plane already found.

The osculating plane at the vertex V_1 (for which $\beta - \gamma = 0$) is found by writing $\alpha = -2$, $\beta = 1$, $\gamma = 1$; its equation is

$$-8x + y + z = 0.$$

The osculating planes at the other two vertices have equations

$$x - 8y + z = 0,$$
$$x + y - 8z = 0.$$

36. Topography of the cubical hyperbola: (ii) the vertices and their osculating planes.

The centre, the central plane, and the axis of the curve have already associated themselves naturally with the skew box.

Next consider the line δ_1 joining the centre to the vertex $V_1\,(0, 3, -3)$; its equations are $x = 0$, $y + z = 0$. The following points on it suggest themselves:

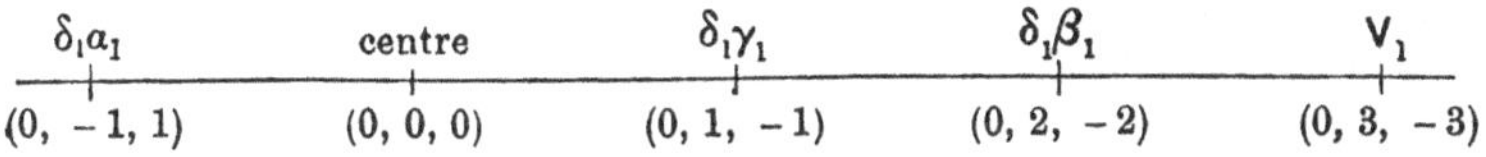

$\delta_1\alpha_1$	centre	$\delta_1\gamma_1$	$\delta_1\beta_1$	V_1
$(0, -1, 1)$	$(0, 0, 0)$	$(0, 1, -1)$	$(0, 2, -2)$	$(0, 3, -3)$

Fig. 5.

The four segments of the line are all equal; also

(i) $\delta_1\alpha_1$ bisects each of the segments $(\alpha_1\beta_2, \alpha_1\beta_3)$, $(\alpha_1\gamma_2, \alpha_1\gamma_3)$.

(ii) $\delta_1\beta_1$ bisects the segment $(\beta_1\alpha_2, \beta_1\alpha_3)$.

(iii) $\delta_1\gamma_1$ bisects the segment $(\gamma_1\alpha_2, \gamma_1\alpha_3)$.

The other two lines δ_2, δ_3 joining the centre to the other two vertices have similar properties.

The osculating plane p_1 at the vertex V_1 is given by $-8x+y+z=0$ and cuts the prism edges β_2, β_3 in the points

$$\beta_2 p_1 (1, 3, 5), \quad \beta_3 p_1 (-1, -5, -3).$$

The point $\beta_1 p_1$ is the same as the point $\beta_1\delta_1$. If p_2 and p_3 are the osculating planes at the other two vertices, the prism edge β_1 passes through the five points

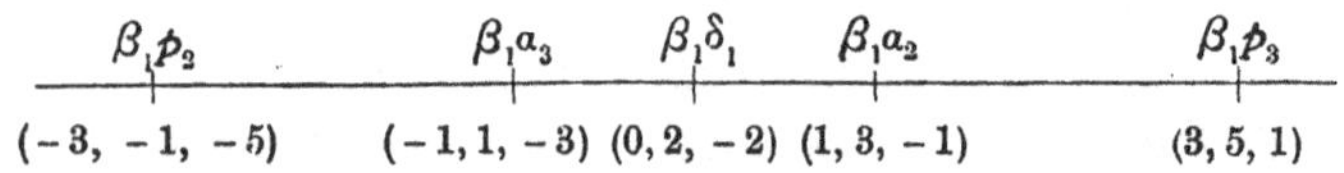

$\beta_1 p_2$	$\beta_1\alpha_3$	$\beta_1\delta_1$	$\beta_1\alpha_2$	$\beta_1 p_3$
$(-3, -1, -5)$	$(-1, 1, -3)$	$(0, 2, -2)$	$(1, 3, -1)$	$(3, 5, 1)$

Fig. 6.

and the three points $\beta_1 p_2$, $\beta_1\delta_1$, $\beta_1 p_3$ make equal segments on β_1.

37. The elements of a cubical hyperbola derived from its asymptotes.

The relations of the cardinal points and lines of the curve to its asymptotes may be summarized as follows:

α_1, α_2, α_3 being the asymptotes, planes are drawn through each of the three lines parallel to the other two; this gives a skew box with three of its edges along α_1, α_2, α_3. If γ_1, γ_2, γ_3 are the edges of the box each meeting two of α_1, α_2, α_3 (so that γ_1 and α_1 are parallel), then γ_1, γ_2, γ_3 are the axes of the circumscribing hyperbolic cylinders; the centre of the box is the centre of the curve, the diagonal of the box which meets none of α_1, α_2, α_3, γ_1, γ_2, γ_3 is the axis of the curve and the plane through the middle points of the six edges along α_1, γ_3, α_2, γ_1, α_3, γ_2 is the central plane of the curve. The planes through α_1, α_2, α_3 parallel to the axis are the asymptotic planes and their intersections determine the three prism edges β_1, β_2, β_3. The line δ_1 joining the centre to the vertex V_1 passes through the middle points of the box edges lying along α_1 and γ_1; the position of V_1 on this line is determined from the fact that the point $\gamma_1\delta_1$ is midway between V_1 and the point $\alpha_1\delta_1$.

38. The form of the cubical hyperbola.

In the equations $x=1-2\theta$, $y=(2-\theta)/\theta$, $z=(\theta+1)/(\theta-1)$, let the parameter θ increase continuously from $-\infty$ to $+\infty$.

When $\theta = -\infty$ the point is the infinity point on α_1; as θ increases to the value -1 the point moves along the curve until it comes to the vertex V_3; and as θ increases to the value 0 the point goes off to infinity parallel to α_2. This gives one branch of the curve. As θ increases from 0 to 1, the point describes another branch which contains the vertex V_1 (for which $\theta = \frac{1}{2}$) and has α_2, α_3 as asymptotes. As θ increases from 1 to ∞, the point describes a third branch which contains the vertex V_2 (for which $\theta = 2$) and has α_3, α_1 as asymptotes.

Each of the branches is associated with one of the variables: the "x-branch" passes through V_1 and has α_2, α_3 as asymptotes, and so on. Let us denote the branches by **x**, **y**, **z**, the portions on one side of the central plane by $\mathbf{x}_+$, $\mathbf{y}_+$, $\mathbf{z}_+$, and those on the other side by $\mathbf{x}_-$, $\mathbf{y}_-$, $\mathbf{z}_-$; then for the point on the curve whose parameter is θ we have

$$x + y + z = -(\theta + 1)(\theta - 2)(2\theta - 1)/\theta(\theta - 1),$$

and the relation between the points of the curve and the values of the parameters is shown by

$$\theta = \underbrace{0 \ldots \tfrac{1}{2}}_{\mathbf{x}_+} \underbrace{\ldots 1}_{\mathbf{x}_-} \underbrace{\ldots 2}_{\mathbf{y}_+} \underbrace{\ldots +\infty}_{\mathbf{y}_-} \quad \underbrace{-\infty \ldots -1}_{\mathbf{z}_+} \underbrace{\ldots 0}_{\mathbf{z}_-}$$

Branches $\mathbf{x}_+$ $\mathbf{x}_-$ $\mathbf{y}_+$ $\mathbf{y}_-$ $\mathbf{z}_+$ $\mathbf{z}_-$

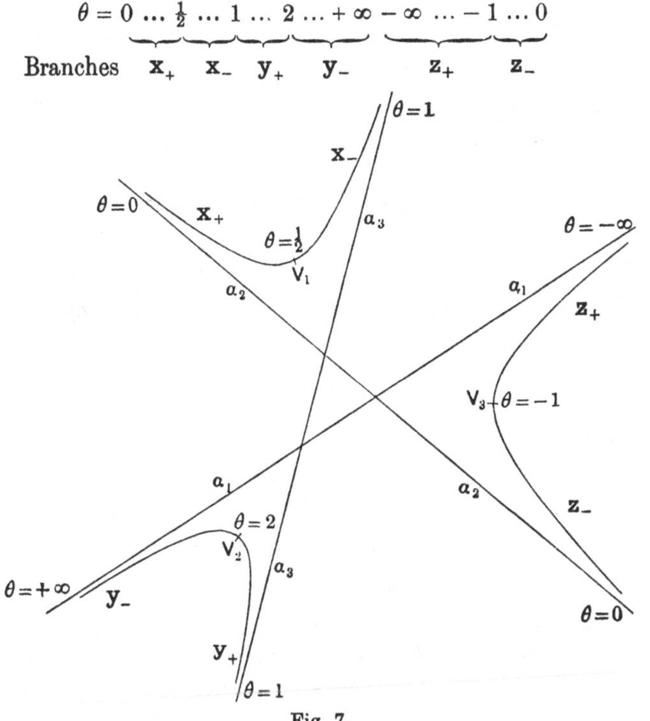

Fig. 7.

39. Arrangement of the branches on the circumscribing cylinders.

We shall show that one sheet of the cylinder $yz + y - z + 3 = 0$ contains the branch **x**, and that the other sheet contains the branches **y**, **z**. The equation of the cylinder is (§ 25)

$$(y-1)(z+1)+4=0,$$

showing $y-1=0$, $z+1=0$ as its asymptotic planes; points whose coordinates give the same signs to $y-1$ and the same signs to $z+1$ lie on the same sheet of the cylinder.

(i) Points on **x** are given by values of θ between 0 and 1: for such points $y\left(=\frac{2-\theta}{\theta}\right)$ lies between 1 and $+\infty$, and $z\left(=\frac{\theta+1}{\theta-1}\right)$ lies between -1 and $-\infty$. So for all points on **x**, $(y-1)$ is positive and $(z+1)$ is negative: all these points lie on one sheet of the cylinder.

(ii) Points on **y** and **z** are given by values of θ between 1 and $+\infty$, and between $-\infty$ and 0; for such points y lies between $-\infty$ and 1, and z lies between -1 and $+\infty$. So $(y-1)$ is negative and $(z+1)$ is positive, and all the points lie on the other sheet of the cylinder.

The arrangement of the branches on the other two circumscribing cylinders follows cyclically.

40. Locus of middle points of chords of the cubical hyperbola.

The coordinates (x, y, z) of the middle point of the chord joining the points with parameters (α, β, γ), $(\alpha', \beta', \gamma')$ are given by

$$2x=\frac{\beta-\gamma}{\alpha}+\frac{\beta'-\gamma'}{\alpha'}$$
$$=-\{(\beta+\gamma)(\beta'-\gamma')+(\beta-\gamma)(\beta'+\gamma')\}/\alpha\alpha',$$

and therefore

$$x=-\frac{\beta\beta'-\gamma\gamma'}{\alpha\alpha'}, \quad y=-\frac{\gamma\gamma'-\alpha\alpha'}{\beta\beta'}, \quad z=-\frac{\alpha\alpha'-\beta\beta'}{\gamma\gamma'}.$$

Hence (§ 31) the locus of the middle points is the cubic surface

$$xyz+x+y+z=0;$$

this surface necessarily contains the curve, since the middle point of an indefinitely short chord lies on the curve.

This surface has nine real generators, namely the three asymptotes, the three cylinder axes, and the three lines δ_1, δ_2, δ_3 (§ 36) joining the centre to the vertices. Each of these nine lines is the locus of the middle points of a system of chords of the cubic and may be called a *diameter* of the curve: the name diameter is however usually restricted to the three lines δ_1, δ_2, δ_3.

41. The diameters of the cubical hyperbola.

(i) The asymptotes.

If the middle point of the chord PP' lies on α_1, we have

$$\alpha\alpha' + \beta\beta' - \gamma\gamma' = 0,$$
$$\alpha\alpha' - \beta\beta' + \gamma\gamma' = 0,$$

which requires $\qquad \alpha\alpha' = 0, \text{ and } \beta\beta' - \gamma\gamma' = 0.$

Taking $\alpha = 0$, then $\beta = -\gamma$, and therefore $\beta' = -\gamma'$ and $\alpha' = 0$, so that the chord PP' coincides with α_1; from the geometrical point of view the asymptote has no claim to be considered as a diameter.

(ii) The cylinder axes.

If the middle point of PP' lies on γ_1, we have the single condition

$$-\alpha\alpha' + \beta\beta' + \gamma\gamma' = 0,$$

which may be written

$$(\beta + \gamma)(\beta' + \gamma') - \beta\beta' - \gamma\gamma' = 0,$$

or

$$\beta\gamma' + \beta'\gamma = 0.$$

The geometry of the curve immediately shows the cylinder axis as a diameter, for any plane through the axis cuts the curve in three points P, Q, P' of which P, Q lie on one sheet of the cylinder and P' on the other sheet, and each of the lines PP', QP' is bisected by the axis.

(iii) The true diameters.

If the middle point of PP' lies on δ_1, then we have the single condition

$$\beta\beta' - \gamma\gamma' = 0.$$

42. Positions of the chords bisected by the diameters.

The ends of the chords bisected by any diameter are companion points in a system of points in involution: for instance the ends of the chords bisected by γ_1 belong to the involution

$$-\alpha\alpha' + \beta\beta' + \gamma\gamma' = 0,$$

or equivalently

$$\frac{1}{\theta} + \frac{1}{\theta'} = 2.$$

This last equation shows immediately how the ends of the chords are situated on the curve; if θ lies between 0 and $\frac{1}{2}$, θ' lies between 0 and $-\infty$, so that one end of the chord is on $\mathbf{x}_+$ and the other on $\mathbf{z}$; if θ lies between $\frac{1}{2}$ and 1, θ' lies between 1 and $+\infty$, so that one end of the chord lies on $\mathbf{x}_-$ and the other on $\mathbf{y}$.

Proceeding similarly with the other involutions we arrive at the following scheme showing the pairing of the branches:

Diameter	*Involution*	*Pairing of branches*	*Double points*
γ_1 $\left.\begin{matrix} y-1=0, \ z+1=0 \end{matrix}\right\}$	$\left.\begin{matrix} -\alpha\alpha'+\beta\beta'+\gamma\gamma'=0 \\ \frac{1}{\theta}+\frac{1}{\theta'}=2 \end{matrix}\right\}$	$\begin{Bmatrix}\mathbf{x}_+\\ \mathbf{z}\end{Bmatrix} \begin{Bmatrix}\mathbf{x}_-\\ \mathbf{y}\end{Bmatrix}$	The infinity points of $\mathbf{x}$
γ_2 $\left.\begin{matrix} z-1=0, \ x+1=0 \end{matrix}\right\}$	$\left.\begin{matrix} \alpha\alpha'-\beta\beta'+\gamma\gamma'=0 \\ \theta+\theta'=2 \end{matrix}\right\}$	$\begin{Bmatrix}\mathbf{y}_+\\ \mathbf{x}\end{Bmatrix} \begin{Bmatrix}\mathbf{y}_-\\ \mathbf{z}\end{Bmatrix}$	The infinity points of $\mathbf{y}$
γ_3 $\left.\begin{matrix} x-1=0, \ y+1=0 \end{matrix}\right\}$	$\left.\begin{matrix} \alpha\alpha'+\beta\beta'-\gamma\gamma'=0 \\ \theta+\theta'=0 \end{matrix}\right\}$	$\begin{Bmatrix}\mathbf{z}_+\\ \mathbf{y}\end{Bmatrix} \begin{Bmatrix}\mathbf{z}_-\\ \mathbf{x}\end{Bmatrix}$	The infinity points of $\mathbf{z}$
δ_1 $\left.\begin{matrix} x=0, \ y+z=0 \end{matrix}\right\}$	$\left.\begin{matrix} \beta\beta'-\gamma\gamma'=0 \\ \theta+\theta'=1 \end{matrix}\right\}$	$\begin{Bmatrix}\mathbf{x}_+\\ \mathbf{x}_-\end{Bmatrix} \begin{Bmatrix}\mathbf{y}_+\\ \mathbf{z}_-\end{Bmatrix} \begin{Bmatrix}\mathbf{y}_-\\ \mathbf{z}_+\end{Bmatrix}$	V_1 and the infinity point on a_1
δ_2 $\left.\begin{matrix} y=0, \ z+x=0 \end{matrix}\right\}$	$\left.\begin{matrix} \gamma\gamma'-\alpha\alpha'=0 \\ \frac{1}{\theta}+\frac{1}{\theta'}=1 \end{matrix}\right\}$	$\begin{Bmatrix}\mathbf{y}_+\\ \mathbf{y}_-\end{Bmatrix} \begin{Bmatrix}\mathbf{z}_+\\ \mathbf{x}_-\end{Bmatrix} \begin{Bmatrix}\mathbf{z}_-\\ \mathbf{x}_+\end{Bmatrix}$	V_2 and the infinity point on a_2
δ_3 $\left.\begin{matrix} z=0, \ x+y=0 \end{matrix}\right\}$	$\left.\begin{matrix} \alpha\alpha'-\beta\beta'=0 \\ \theta\theta'=1 \end{matrix}\right\}$	$\begin{Bmatrix}\mathbf{z}_+\\ \mathbf{z}_-\end{Bmatrix} \begin{Bmatrix}\mathbf{x}_+\\ \mathbf{y}_-\end{Bmatrix} \begin{Bmatrix}\mathbf{x}_-\\ \mathbf{y}_+\end{Bmatrix}$	V_3 and the infinity point on a_3

43. Quadrics circumscribing the cubical hyperbola.

The three circumscribing cylinders being given by

$$u \equiv yz + y - z + 3 = 0,$$

$$v \equiv zx + z - x + 3 = 0,$$

$$w \equiv xy + x - y + 3 = 0,$$

any circumscribing quadric is given (§ 4, (ii)) by the equation

$$s \equiv Au + Bv + Cw = 0,$$

where A, B, C are arbitrary constants. If P, P' are two points of the cubic, s will contain the chord PP' as a generator if it contains the infinity point of PP', since it necessarily contains the points P, P' themselves. This infinity point has coordinates proportional to

$$\left(\frac{\beta-\gamma}{\alpha}-\frac{\beta'-\gamma'}{\alpha'},\ \frac{\gamma-\alpha}{\beta}-\frac{\gamma'-\alpha'}{\beta'},\ \frac{\alpha-\beta}{\gamma}-\frac{\alpha'-\beta'}{\gamma'}\right),$$

or to
$$\left(\frac{1}{\alpha\alpha'}, \frac{1}{\beta\beta'}, \frac{1}{\gamma\gamma'}\right),$$
since
$$\beta\gamma' - \beta'\gamma = \gamma\alpha' - \gamma'\alpha = \alpha\beta' - \alpha'\beta.$$

So s contains PP' as a generator if
$$A\alpha\alpha' + B\beta\beta' + C\gamma\gamma' = 0.$$
This equation gives the involution of points associated (§ 10) with the quadric.

The coordinates of the centre of the quadric are given by
$$B(z-1)+C(y+1)=0, \quad C(x-1)+A(z+1)=0, \quad A(y-1)+B(x+1)=0,$$
and therefore the locus of centres of quadrics circumscribing the curve is the cubic surface
$$(x+1)(y+1)(z+1)+(x-1)(y-1)(z-1)=0,$$
or
$$xyz + x + y + z = 0.$$

We saw in § 31 that this surface circumscribes the cubic curve and in § 40 that it is the locus of the middle points of all chords of the curve.

44. The unique chord through any point.

The circumscribing quadric $Au + Bv + Cw = 0$ passes through the point P_1 if $Au_{11} + Bv_{11} + Cw_{11} = 0$, where $u_{11} \equiv y_1 z_1 + y_1 - z_1 + 3$, etc., and has the unique chord through P_1 (§ 10) as a generator. By varying A, B, C we obtain a system of circumscribing quadrics having this unique chord as a common generator. If (α, β, γ), $(\alpha', \beta', \gamma')$ are the ends of this chord, then (§ 43) $A\alpha\alpha' + B\beta\beta' + C\gamma\gamma' = 0$ for all values of A, B, C satisfying the equation $Au_{11} + Bv_{11} + Cw_{11} = 0$, and therefore $\alpha\alpha' : \beta\beta' : \gamma\gamma' = u_{11} : v_{11} : w_{11}$. These equations with $\alpha + \beta + \gamma = 0$, $\alpha' + \beta' + \gamma' = 0$ determine the parameters of the ends of the chord.

(i) If the point (α, β, γ) is fixed, and $(\alpha', \beta', \gamma')$ is variable, then
$$\frac{u_{11}}{\alpha} + \frac{v_{11}}{\beta} + \frac{w_{11}}{\gamma} = 0$$
is the condition satisfied by the coordinates of a point P_1 lying on any chord through (α, β, γ). So the quadric cone projecting the curve from the point (α, β, γ) of the curve is given by the equation
$$\frac{u}{\alpha} + \frac{v}{\beta} + \frac{w}{\gamma} = 0.$$

(ii) If the ends of the chord are both coincident with the point (α, β, γ), the coordinates of any point on the tangent at that point satisfy the equations

$$\frac{u}{\alpha^2}=\frac{v}{\beta^2}=\frac{w}{\gamma^2},$$

and therefore the tangents generate (§ 13) the developable surface

$$\sqrt{u}+\sqrt{v}+\sqrt{w}=0,$$

or
$$u^2+v^2+w^2-2vw-2wu-2uv=0.$$

(iii) The direction of the chord PP' is given by $\left(\frac{1}{\alpha\alpha'}, \frac{1}{\beta\beta'}, \frac{1}{\gamma\gamma'}\right)$, that is by $\left(\frac{1}{u_{11}}, \frac{1}{v_{11}}, \frac{1}{w_{11}}\right)$. The unique chord through the centre is therefore $x=y=z$, the axis of the curve: since three real osculating planes (namely those at the vertices) pass through the centre, the ends of the chord are imaginary. Their parameters are given by

$$\alpha\alpha'=\beta\beta'=\gamma\gamma',$$

whence
$$\left.\begin{array}{l}\alpha\ :\beta\ :\gamma\ =1:\omega\ :\omega^2\\ \alpha':\beta':\gamma'=1:\omega^2:\omega\end{array}\right\},$$

where ω is an imaginary cube root of unity.

The equations of the osculating planes at these imaginary points are (§ 35) found to be

$$x+y+z\pm 3i\sqrt{3}=0;$$

they are a conjugate imaginary pair of parallel planes having the central plane of the curve midway between them (see § 11).

This pair of imaginary points is common to each of the involutions of § 42 with equations

$$\beta\beta'-\gamma\gamma'=0, \quad \gamma\gamma'-\alpha\alpha'=0, \quad \alpha\alpha'-\beta\beta'=0.$$

(iv) P and P' being the ends of the unique chord through P_1 and P_2 being the harmonic conjugate of P_1 with respect to P, P', the points P_1 and P_2 are said to be conjugate with regard to the twisted cubic. Any pair of conjugate points are also conjugate with regard to all quadrics circumscribing the cubic: specially the centre and the infinity point of the axis are conjugate.

Some of the properties of conjugate points have been discussed by Reye*.

* *Geometrie der Lage*, Bd. II.

45. Special involutions of points on the curve.

Consider first the six involutions (§ 42) which determine the chords bisected by the diameters.

(i) The ends of the chords bisected by the true diameter δ_1 are corresponding points of the involution

$$\beta\beta' - \gamma\gamma' = 0.$$

The chords form one system of generators of the quadric (§ 43) $v - w = 0$, or

$$x(y - z) + 2x - y - z = 0;$$

this is a hyperbolic paraboloid. The bisected chords are all parallel to the asymptotic plane $y - z + 2 = 0$ of the curve; among the generators of the other system (all parallel to $x = 0$) are

$$x = 0, \quad y + z = 0;$$

$$x + 1 = 0, \quad y + 1 = 0;$$

$$x - 1 = 0, \quad z - 1 = 0.$$

The first of these is the diameter δ_1 bisecting the chords; the other two are two edges of the skew box (§ 28) which now associate themselves with the curve for the first time. The intercept made by them on any chord of the system we are considering is also bisected by the diameter δ_1.

Similar properties hold for the chords bisected by the diameters δ_2, δ_3; all the twelve edges of the skew box have now been related to the curve.

(ii) The chords bisected by the cylinder axis γ_1 are determined by the involution

$$-\alpha\alpha' + \beta\beta' + \gamma\gamma' = 0,$$

and generate the quadric (§ 43)

$$-u + v + w = 0,$$

or

$$-yz + zx + xy - 2y + 2z + 3 = 0.$$

This is a hyperboloid with its centre at the point $\beta_1\delta_1$.

(iii) Consider also the involution $\alpha\alpha' + \beta\beta' + \gamma\gamma' = 0$: its double points given by $\alpha^2 + \beta^2 + \gamma^2 = 0$ are the imaginary ends of the axis: the chords generate the quadric (§ 43) $u + v + w = 0$, or

$$yz + zx + xy + 9 = 0.$$

The interest of this involution will appear in § 49.

46. Polarization with respect to the screw containing the tangents.

So far we have regarded the curve as the locus of its points, and not as being osculated by its osculating planes; correlative results are immediately derivable by polarization. Any chord is polarized into the common line of the two osculating planes at its ends; the polarized equation of a circumscribing quadric represents an inscribed quadric (§ 9); specially the quadric cone projecting the curve from any point P of itself is polarized into a conic lying in the osculating plane at P and touching all the osculating planes of the curve.

The pole of the plane p_1 with respect to the screw (§ 30)

$$2(l+m+n)+L+M+N=0$$

has for its equation (cf. § 32)

$$(Y_1-Z_1+\tfrac{1}{2})X+(Z_1-X_1+\tfrac{1}{2})Y+(X_1-Y_1+\tfrac{1}{2})Z-\tfrac{1}{2}(X_1+Y_1+Z_1)=0,$$

and the process of polarization is therefore equivalent to the transformation from point coordinates (x, y, z) to plane coordinates (X, Y, Z) given by the equations

$$x=-\{2(Y-Z)+1\}/(X+Y+Z),$$
$$y=-\{2(Z-X)+1\}/(X+Y+Z),$$
$$z=-\{2(X-Y)+1\}/(X+Y+Z).$$

The pole of the plane at infinity is the point at infinity on the axis of the curve: it follows that all the diameters of the screw are parallel to the axis of the curve; the asymptotes $\alpha_1, \alpha_2, \alpha_3$ and the diameters $\delta_1, \delta_2, \delta_3$ are rays of the screw. The prism edge β_1 meets those rays of the screw which are parallel to $x=0$; the polar line of the axis of the curve is the line at infinity in the central plane. A formula for the pitch of the screw is found in § 53.

47. Quadrics inscribed in the cubical hyperbola.

Since $u\equiv yz+y-z+3=0$ is the equation of the cylinder projecting the curve from the infinity point on the asymptote α_1, the equation of the conic envelope of osculating planes lying in the asymptotic plane $y-z+2=0$ is

$$(2Z-2X+1)(2X-2Y+1)-(X+Y+Z)(2Z-2X+1)$$
$$+(X+Y+Z)(2X-2Y+1)+3(X+Y+Z)^2=0.$$

This can be written in the form

$$U\equiv(-Y+Z+1)^2+3X(X+4Y+4Z)=0,$$

showing the centre of the conic as the point $(\alpha_1\delta_1)$ with equation $-Y+Z+1=0$, and the two real infinity points with equations

$$X=0, \quad X+4Y+4Z=0;$$

the conic is a hyperbola in the plane $y-z+2=0$.

The conics inscribed in the developable and lying in the other asymptotic planes are given similarly by the equations

$$V \equiv (-Z+X+1)^2+3Y(4X+Y+4Z)=0,$$

with centre $\alpha_2\delta_2$;

$$W \equiv (-X+Y+1)^2+3Z(4X+4Y+Z)=0,$$

with centre $\alpha_3\delta_3$.

Any quadric inscribed in the curve is given (§ 9) by the equation

$$AU+BV+CW=0,$$

where A, B, C are constants, one set of generators being the intersections of osculating planes at corresponding points of the involution determined by the equation

$$A\alpha\alpha'+B\beta\beta'+C\gamma\gamma'=0.$$

Specially the conic inscribed in the developable and lying in the osculating plane at the point (α, β, γ) has for its equation (§ 44)

$$\frac{U}{\alpha}+\frac{V}{\beta}+\frac{W}{\gamma}=0.$$

The equation of the centre of the quadric given by

$$AU+BV+CW=0$$

is $\quad A(-Y+Z+1)+B(-Z+X+1)+C(-X+Y+1)=0.$

Hence the centres of all inscribed quadrics lie in the plane through the three points $(0, -1, 1)$, $(1, 0, -1)$, $(-1, 1, 0)$; this plane is the central plane.

In order to determine the unique line-in-two-planes lying in any plane p_1 we need only polarize the result of § 44; if (α, β, γ), $(\alpha', \beta', \gamma')$ are the parameters of the points whose osculating planes are collinear with p_1, then

$$\frac{\alpha\alpha'}{U_{11}}=\frac{\beta\beta'}{V_{11}}=\frac{\gamma\gamma'}{W_{11}},$$

where $\quad U_{11} \equiv (-Y_1+Z_1+1)^2+3X_1(X_1+4Y_1+4Z_1)$, etc.

48. Locus of centres of inscribed conics.

The equation of any inscribed conic is

$$\frac{U}{\alpha}+\frac{V}{\beta}+\frac{W}{\gamma}=0,$$

where
$$\alpha + \beta + \gamma = 0,$$
and the equation of its centre is
$$\frac{1}{\alpha}(-Y+Z+1)+\frac{1}{\beta}(-Z+X+1)+\frac{1}{\gamma}(-X+Y+1)=0.$$

The locus is to be found from the condition that the equations
$$(-Y+Z+1)\frac{d\alpha}{\alpha^2}+(-Z+X+1)\frac{d\beta}{\beta^2}+(-X+Y+1)\frac{d\gamma}{\gamma^2}=0,$$
and
$$d\alpha + d\beta + d\gamma = 0,$$
are simultaneously true for all values of the increments $d\alpha : d\beta : d\gamma$. This requires
$$\frac{-Y+Z+1}{\alpha^2}=\frac{-Z+X+1}{\beta^2}=\frac{-X+Y+1}{\gamma^2};$$
the equation of the locus is therefore
$$\sqrt{-Y+Z+1}+\sqrt{-Z+X+1}+\sqrt{-X+Y+1}=0,$$
or, in its expanded form,
$$4(X^2+Y^2+Z^2)-4(YZ+ZX+XY)-3=0.$$
This equation represents a conic in the central plane $x+y+z=0$. Locus equations of the conic may be deduced from this or found independently as follows:

The coordinates of the centre of $\frac{U}{\alpha}+\frac{V}{\beta}+\frac{W}{\gamma}=0$ are given by
$$x : y : z : 1 = \frac{1}{\beta}-\frac{1}{\gamma} : \frac{1}{\gamma}-\frac{1}{\alpha} : \frac{1}{\alpha}-\frac{1}{\beta} : \frac{1}{\alpha}+\frac{1}{\beta}+\frac{1}{\gamma}.$$

So $x+y+z=0$, showing the locus as lying in the central plane; also
$$\frac{1}{y-z-1}+\frac{1}{z-x-1}+\frac{1}{x-y-1}=0;$$
using the equation $x+y+z=0$, this reduces to
$$yz+zx+xy+1=0.$$
This equation represents a quadric containing the asymptotes and cylinder axes as generators: hence the locus of centres of inscribed conics is the section, by the central plane, of the quadric which contains the three asymptotes and the three cylinder axes.

The conic passes through the six points
$$(0,\ \pm 1,\ \mp 1),\quad (\mp 1,\ 0,\ \pm 1),\quad (\pm 1,\ 0,\ \mp 1),$$
where the central plane is met by the asymptotes and cylinder axes.

49. The conjugate cubical hyperbola.

Associated with any cubical hyperbola is another cubical hyperbola which is the reflexion of the first in its own centre: two such cubical hyperbolas are called conjugate.

The parametric equations of the conjugate hyperbola are found by changing the signs of the coordinates x, y, z; they are

$$x=-\frac{\beta-\gamma}{\alpha},\quad y=-\frac{\gamma-\alpha}{\beta},\quad z=-\frac{\alpha-\beta}{\gamma};\quad \alpha+\beta+\gamma=0.$$

They may also be obtained by interchanging the parametric expressions for any two of the coordinates; for writing

$$x=\frac{\beta-\gamma}{\alpha},\quad z=\frac{\gamma-\alpha}{\beta},\quad y=\frac{\alpha-\beta}{\gamma};\quad \alpha+\beta+\gamma=0,$$

and $$\alpha=\alpha',\quad \beta=\gamma',\quad \beta'=\gamma,$$

then $$x=-\frac{\beta'-\gamma'}{\alpha'},\quad y=-\frac{\gamma'-\alpha'}{\beta'},\quad z=-\frac{\alpha'-\beta'}{\gamma'};\quad \alpha'+\beta'+\gamma'=0.$$

A cubical hyperbola and its conjugate have in common the centre, the central plane, the central axis, the true diameters δ_1, δ_2, δ_3, the osculating planes at the vertices, the cubic surface locus of the middle points of chords (which is also the locus of the centres of circumscribing quadrics), and the conic locus of centres of inscribed conics. The asymptotes of each curve are the axes of the cylinders circumscribing the other.

The cubical hyperbola and its conjugate constitute the complete intersection of the surfaces

$$xyz+x+y+z=0,$$
$$yz+zx+xy+9=0.$$

We found in § 40 that the middle point of the chord joining the points (α, β, γ), $(\alpha', \beta', \gamma')$ has coordinates

$$x=-\frac{\beta\beta'-\gamma\gamma'}{\alpha\alpha'},\quad y=-\frac{\gamma\gamma'-\alpha\alpha'}{\beta\beta'},\quad z=-\frac{\alpha\alpha'-\beta\beta'}{\gamma\gamma'}.$$

Let the ends of the chord be corresponding points of the elliptic involution

$$\alpha\alpha'+\beta\beta'+\gamma\gamma'=0,$$

then the locus of the middle points is the conjugate cubical hyperbola, for $(\alpha\alpha', \beta\beta', \gamma\gamma')$ can be taken as parameters in place of (α, β, γ). The involution is determined from the property that its double points are the imaginary ends of the chord lying along the axis (§ 44 (iii)).

50. Any plane cuts the cubical hyperbola and its asymptotes in two triangles with the same centroid.

The plane $Ax + By + Cz + D = 0$ cuts the curve in three points whose parameters θ_1, θ_2, θ_3 are roots of the cubic

$$A(1-2\theta) + B\frac{2-\theta}{\theta} + C\frac{\theta+1}{\theta-1} + D = 0,$$

and so
$$\theta_1 + \theta_2 + \theta_3 = \frac{3A - B + C + D}{2A}.$$

Hence if (x, y, z) are the coordinates of the centroid

$$3x = 3 - 2(\theta_1 + \theta_2 + \theta_3)$$
$$= \frac{B - C - D}{A},$$

and similarly

$$3y = \frac{C - A - D}{B}, \quad 3z = \frac{A - B - D}{C}.$$

The plane meets the asymptotes in the points

$$\left(\frac{B-C-D}{A}, -1, 1\right), \quad \left(1, \frac{C-A-D}{B}, -1\right), \quad \left(-1, 1, \frac{A-B-D}{C}\right),$$

whose centroid is the point

$$\left(\frac{B-C-D}{3A}, \frac{C-A-D}{3B}, \frac{A-B-D}{3C}\right),$$

and so the two centroids coincide.

The result is analogous to the property of the plane hyperbola that the intercepts on any line by the curve and by the asymptotes have the same middle point.

Specially the osculating plane at any point P of the curve cuts the asymptotes in three points whose centroid is at P.

If we take a system of planes parallel to $Ax + By + Cz = 0$, the centroids of the triangles in which they cut the curve lie on the line given by

$$Ax - \frac{B-C}{3} = By - \frac{C-A}{3} = Cz - \frac{A-B}{3}.$$

The point

$$\left(\frac{B-C-D}{3A}, \frac{C-A-D}{3B}, \frac{A-B-D}{3C}\right)$$

lies on the surface

$$3yz - y + z + 1 = 0,$$

if
$$D^2 = A^2 + B^2 + C^2 - 2BC - 2CA - 2AB.$$

Selecting from the parallel system the two planes given by these values of D, the corresponding centroids lie on each of the cylinders

$$3yz - y + z + 1 = 0,$$
$$3zx - z + x + 1 = 0,$$
$$3xy - x + y + 1 = 0.$$

The centroid locus is therefore a chord of the twisted cubic which is the common curve of these three cylinders. This twisted cubic is given parametrically by the equations

$$-3x = \frac{\beta - \gamma}{\alpha}, \quad -3y = \frac{\gamma - \alpha}{\beta}, \quad -3z = \frac{\alpha - \beta}{\gamma}; \quad \alpha + \beta + \gamma = 0;$$

it is similar to, similarly situated and concentric with the conjugate cubical hyperbola and is one-third of its linear dimensions.

The theorem may be generalized in the following form*: "a variable plane is drawn through a fixed line λ cutting a twisted cubic in three points; then the locus of the harmonic poles of λ with respect to these triads of points is a straight line." This may be proved otherwise by the method of § 12.

51. The locus of the centroid of the points in which three fixed generators of a quadric are met by a variable generator is a twisted cubic.

Taking the equations of the three fixed generators to be

$$\left.\begin{array}{l} y+1=0 \\ z-1=0 \end{array}\right\}, \quad \left.\begin{array}{l} z+1=0 \\ x-1=0 \end{array}\right\}, \quad \left.\begin{array}{l} x+1=0 \\ y-1=0 \end{array}\right\},$$

the quadric containing them is given by

$$yz + zx + xy + 1 = 0.$$

Any generator of the opposite system has equations

$$\alpha x - \gamma z + \beta = 0,$$
$$-\alpha x + \beta y + \gamma = 0,$$

where α, β, γ are parameters such that $\alpha + \beta + \gamma = 0$.

This generator meets the fixed generators in the points with coordinates

$$\left(-\frac{\beta - \gamma}{\alpha}, -1, 1\right), \quad \left(1, -\frac{\gamma - \alpha}{\beta}, -1\right), \quad \left(1, -1, -\frac{\alpha - \beta}{\gamma}\right),$$

* A. C. Dixon, *Quarterly Journal of Mathematics*, Vol. XXIV, 1890.

and therefore the coordinates of the centroid of the three points are given by

$$-3x=\frac{\beta-\gamma}{\alpha},\quad -3y=\frac{\gamma-\alpha}{\beta},\quad -3z=\frac{\alpha-\beta}{\gamma}.$$

The locus of the centroid is therefore the twisted cubic of § 50.

52. Another method of generating the cubical hyperbola.

Three lines γ_1, γ_2, γ_3 being taken arbitrarily, a variable line λ meeting all three is taken (so that λ is a generator of the quadric which has γ_1, γ_2, γ_3 as generators); through the point $\lambda\gamma_1$ is drawn a plane parallel to γ_2 and γ_3, through the point $\lambda\gamma_2$ a plane parallel to γ_3 and γ_1, and through the point $\lambda\gamma_3$ a plane parallel to γ_1 and γ_2. If P is the common point of these three planes, we shall show that the locus of P, as λ varies, is a cubical hyperbola, the given lines γ_1, γ_2, γ_3 being the axes of the cylinders circumscribing the cubic.

To prove this, take the lines γ_1, γ_2, γ_3 to be given by the equations

$$\left.\begin{matrix} y-1=0 \\ z+1=0 \end{matrix}\right\},\quad \left.\begin{matrix} z-1=0 \\ x+1=0 \end{matrix}\right\},\quad \left.\begin{matrix} x-1=0 \\ y+1=0 \end{matrix}\right\},$$

then the three points, one on each of these lines, with coordinates

$$(x', 1, -1),\quad (-1, y', 1),\quad (1, -1, z')$$

are collinear, if

$$\left\|\begin{matrix} x' & 1 & -1 & 1 \\ -1 & y' & 1 & 1 \\ 1 & -1 & z' & 1 \end{matrix}\right\|=0,$$

and therefore

$$y'z'+y'-z'+3=0,$$
$$z'x'+z'-x'+3=0,$$
$$x'y'+x'-y'+3=0.$$

The point of concurrence of the three planes $x-x'=0$, $y-y'=0$, $z-z'=0$ is the point (x', y', z'), and its locus is the cubical hyperbola of § 28.

The result may be stated reversely: through any point P of a cubical hyperbola are drawn three planes each parallel to two of the asymptotes; these three planes meet the corresponding axes of the circumscribing cylinders in three collinear points.

So, associated with any point P (α, β, γ) of the curve is a generator λ of the quadric $yz+zx+xy+1=0$, with coordinates

$$(\beta\gamma,\ \gamma\alpha,\ \alpha\beta,\ \alpha^2,\ \beta^2,\ \gamma^2).$$

The verification of this and the following results is left to the reader.

(i) λ is a ray of the screw containing the tangents.

(ii) The plane $P\lambda$ has equation $\alpha^2x+\beta^2y+\gamma^2z=0$ and touches the asymptotic cone $yz+zx+xy=0$ of the quadric along the generator given by $\alpha x=\beta y=\gamma z$.

(iii) If Q is the point in which λ meets the central plane $x+y+z=0$, then Q lies on the osculating plane at P and is the centre of the inscribed conic (§ 48) lying in that plane. Q is also the pole of the plane $P\lambda$ with respect to the screw containing the tangents, and PQ is a ray of the screw.

53. Pitch of the screw containing the tangents*.

The axis of the screw is parallel to the axis of the curve: its pitch ϖ is given by the formula

$$\varpi=q\tan\theta,$$

where q is the length of the common normal of the screw axis and any ray, and θ is the angle between that ray and the screw axis. The asymptotes themselves are rays of the screw and so, if q_1, q_2, q_3 are the common normals of the asymptotes and the screw axis and $\theta_1, \theta_2, \theta_3$ the angles between the asymptotes and the screw axis,

$$\varpi=q_1\tan\theta_1=q_2\tan\theta_2=q_3\tan\theta_3.$$

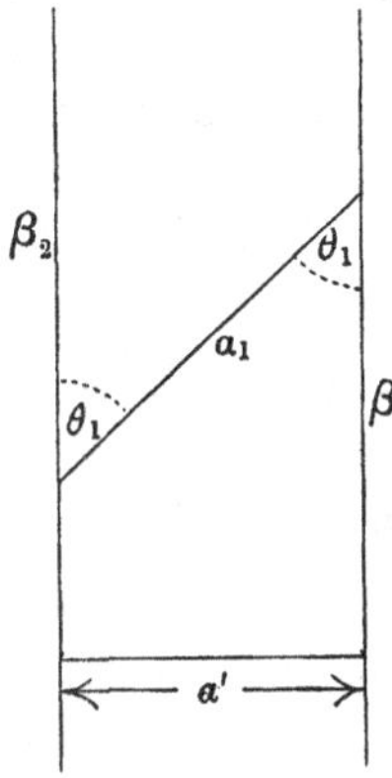

Let any plane p perpendicular to the edges of the triangular prism of asymptotic planes cut it in a triangle of sides a', b', c' and area Δ, and let l_1, l_2, l_3 be the lengths intercepted on the asymptotes by the edges of the prism, so that

Fig. 8.

$$a'=l_1\sin\theta_1,\quad b'=l_2\sin\theta_2,\quad c'=l_3\sin\theta_3.$$

Then
$$\varpi=q_1\tan\theta_1=\frac{q_1a'}{l_1\cos\theta_1},$$

so
$$\varpi l_1\cos\theta_1=a'q_1,\quad \varpi l_2\cos\theta_2=b'q_2,\quad \varpi l_3\cos\theta_3=c'q_3.$$

Now
$$a'q_1+b'q_2+c'q_3=2\Delta,$$

and
$$l_1\cos\theta_1+l_2\cos\theta_2+l_3\cos\theta_3=3l,$$

* See Heinrichs, *Zeitschrift für Math. u. Phys.* XXXIX, 1894.

where l is the length intercepted on each prism edge by the two asymptotes which meet it. Hence

$$\varpi = \frac{2\Delta}{3l}.$$

54. Relations between a cubical hyperbola and the imaginary circle at infinity.

The metrical relations considered up to this point (excluding § 53) have involved only the plane at infinity and not the imaginary circle at infinity.

Most of the properties of the curve involving rectangularity and circularity have been established by their authors by methods of pure geometry. In order to deal with some of these analytically, we use oblique Cartesian coordinates; the preceding formulae are valid if the point coordinates (x, y, z) are replaced by $(x/a, y/b, z/c)$, and the plane coordinates (X, Y, Z) by (aX, bY, cZ), where $2a$, $2b$, $2c$ are the edges of the skew box.

The parametric equations of the curve are now

$$x = a\frac{\beta-\gamma}{\alpha}, \quad y = b\frac{\gamma-\alpha}{\beta}, \quad z = c\frac{\alpha-\beta}{\gamma}, \text{ with } \alpha+\beta+\gamma=0;$$

the equation $yz + cy - bz + 3bc = 0$ represents one of the circumscribing cylinders, and so on.

If the cosines of the angles between the asymptotes are ϵ_1, ϵ_2, ϵ_3, the equation of the isotropic cone with its vertex at the origin is

$$x^2 + y^2 + z^2 + 2\epsilon_1 yz + 2\epsilon_2 zx + 2\epsilon_3 xy = 0,$$

and the equation of the imaginary circle at infinity is

$$(1-\epsilon_1^2)X^2 + (1-\epsilon_2^2)Y^2 + (1-\epsilon_3^2)Z^2 + 2(\epsilon_2\epsilon_3 - \epsilon_1)YZ$$
$$+ 2(\epsilon_3\epsilon_1 - \epsilon_2)ZX + 2(\epsilon_1\epsilon_2 - \epsilon_3)XY = 0.$$

55. The four quadrics of revolution circumscribing the cubical hyperbola.

If the quadric given (§ 43) by the equation

$$A(yz + cy - bz + 3bc) + B(zx + \ldots) + C(xy + \ldots) = 0$$

is one of revolution, a parameter k can be found so that

$$x^2 + y^2 + z^2 + 2(\epsilon_1 + Ak)yz + 2(\epsilon_2 + Bk)zx + 2(\epsilon_3 + Bk)xy$$

is a perfect square. The form of the equation shows that the perfect square is one of the four $(x \pm y \pm z)^2$.

(i) If the quadric has its cyclic planes parallel to $x+y+z=0$, then

$$1=\epsilon_1+Ak=\epsilon_2+Bk=\epsilon_3+Ck,$$

and so $$A:B:C=1-\epsilon_1:1-\epsilon_2:1-\epsilon_3.$$

(ii) Corresponding to a cyclic plane $-x+y+z=0$, we have

$$1=\epsilon_1+Ak=-(\epsilon_2+Bk)=-(\epsilon_3+Ck),$$

and so $$A:B:C=1-\epsilon_1:-(1+\epsilon_2):-(1+\epsilon_3).$$

(iii) If $x-y+z=0$ is a cyclic plane,

$$A:B:C=-(1+\epsilon_1):1-\epsilon_2:-(1+\epsilon_3).$$

(iv) If $x+y-z=0$ is a cyclic plane,

$$A:B:C=-(1+\epsilon_1):-(1+\epsilon_2):1-\epsilon_3.$$

There are therefore four real quadrics of revolution circumscribing the curve and their cyclic planes are parallel to the four planes

$$x\pm y\pm z=0.$$

The lines bisecting the angles between the coordinate axes are given by the three pairs of equations

$$x=0,\ y\pm z=0;\quad y=0,\ z\pm x=0;\quad z=0,\ x\pm y=0.$$

Through any point draw three lines parallel to the asymptotes of the curve and draw lines (perpendicular in pairs) bisecting the angles between them. These six lines lie by threes in four planes which are parallel to the cyclic planes of the four revolution quadrics circumscribing the curve.

These results due to Cremona* may be obtained synthetically as follows. Any quadric circumscribing the curve is uniquely determined by its infinity section, since it is to contain two points of that section other than the three lying on the curve. The problem is therefore reduced to the determination of any cone homothetic with its asymptotic cone from the conditions that the cone is circular and has three generators parallel to the asymptotes. Stated in terms of spherical geometry the problem is reciprocal to that of finding the circles touching the sides of a spherical triangle. There are four solutions of this problem and the geometry of the figure shows further that the plane parallel to two of the axes of revolution is perpendicular to the plane parallel to the other two axes of revolution.

From the existence of four real revolution quadrics circumscribing the curve Professor Dixon has established a series of properties having

* *Crelle*, LXIII.

some analogy with those of conics. His paper will be found in Vol. XXIV of the *Quarterly Journal of Mathematics* (1890).

56. Equiangular hyperboloids inscribed in the cubical hyperbola.

By the term equiangular hyperboloid is to be understood a hyperboloid whose asymptotic cone possesses an infinite number of triads of mutually perpendicular tangent planes: the director sphere of any such quadric is a point sphere.

First consider the system of quadrics inscribed in the curve: they are given by the equation (§ 47)

$$AU+BV+CW=0,$$

where

$$U\equiv(-bY+cZ+1)^2+3aX(aX+4bY+4cZ)=0, \text{ etc.}$$

and A, B, C are parameters. The locus equations of the director spheres of these quadrics are linear in A, B, C and hence these director spheres have two common points, say P, Q; the centres of the ∞^2 quadrics lie on the central plane and therefore PQ is bisected perpendicularly by the central plane. Among the ∞^2 quadrics there are ∞^1 which are equiangular: their director spheres are point spheres lying on the central plane. These point spheres pass through the points P, Q and therefore lie on a circle in the central plane, the circle being imaginary if P and Q are real and conversely. It follows that the locus of the centres of inscribed equiangular hyperboloids is a circle lying in the central plane. This circle and the conic locus of centres of inscribed conics have four common points; if an equiangular hyperboloid degenerates into a conic, the conic is a rectangular hyperbola. Hence four of the inscribed conics are rectangular hyperbolas and their centres are concyclic*.

57. Loci connected with a system of parallel planes.

It has been proved (§ 50) that the centroids of the triangles in which the curve is cut by a system of parallel planes are collinear. We shall now show that the centres of the circles circumscribing these triangles are collinear.

First it will be proved that a quadric can always be found to circumscribe the curve and to have one set of cyclic planes parallel to a given plane.

* See Krüger, *Zeitschrift für Mathematik und Physik*, XXXVIII, 1893.

If the quadric

$$Ayz + Bzx + Cxy + \dots = 0$$

has a set of cyclic planes parallel to the plane given by

$$X_1 x + Y_1 y + Z_1 z = 0,$$

a parameter k can be found such that

$$x^2 + y^2 + z^2 + 2(\epsilon_1 + Ak)yz + 2(\epsilon_2 + Bk)zx + 2(\epsilon_3 + Ck)xy$$
$$\equiv (X_1 x + Y_1 y + Z_1 z)\left(\frac{x}{X_1} + \frac{y}{Y_1} + \frac{z}{Z_1}\right),$$

and so $$2(\epsilon_1 + Ak) = \frac{Y_1}{Z_1} + \frac{Z_1}{Y_1}, \text{ etc.}$$

and therefore

$$A : B : C = \frac{Y_1}{Z_1} + \frac{Z_1}{Y_1} - 2\epsilon_1 : \frac{Z_1}{X_1} + \frac{X_1}{Z_1} - 2\epsilon_2 : \frac{X_1}{Y_1} + \frac{Y_1}{X_1} - 2\epsilon_3.$$

The result required now follows: any plane parallel to the plane $X_1 x + Y_1 y + Z_1 z = 0$ cuts the curve in three points lying on a circular section of the quadric and therefore the circumcentre locus is the diameter of the quadric conjugate to these circular sections. Stated synthetically, the infinity section of the quadric, and therefore the quadric itself, is uniquely determined as the conic through the three infinity points of the curve and two given points of the imaginary circle at infinity.

The line joining the centroid G and the circumcentre O of a triangle is the "Euler line" of the triangle: it also passes through the orthocentre P and the nine points centre N; N bisects OP, and $2OG = GP$.

Fig. 9.

The Euler lines of the set of triangles meet the centroid locus and the circumcentre locus and are parallel to a fixed plane; they therefore generate a hyperbolic paraboloid. For all positions of the Euler line the ratios $OG : GN : NP$ are constant and therefore each of the points N, P traces out a generator of the hyperbolic paraboloid. It follows that the locus of the orthocentres of the system of triangles is a straight line and that the locus of the nine points centres of the system of triangles is a straight line.

A full discussion of these and allied loci has been given by Dixon*: the question has also been treated by Krüger† who proves further that

(i) any chord of the curve is the orthocentre locus of four sets of triangles;

* *Quarterly Journal of Mathematics*, XXIV, 1890.

† *Zeitschrift für Math. u. Phys.* XL, 1895.

(ii) any point in space is the centre of ten circles each of which cuts the curve in three points.

58. Special metrical cubical hyperbolas.

In conclusion we shall consider briefly some metrical properties of special cubical hyperbolas.

The cubical hyperbola will be called *equilateral* when the edges of the skew box derived as in § 37 from its asymptotes are equal.

If we take the length of each edge to be 2, the formulae and results of §§ 28—52 are all available on the understanding that (x, y, z) are oblique Cartesian coordinates. The dimensions of the curve are easily calculated; for instance, we find

(i) the intercept on any asymptote by the edges of the prism of asymptotic planes is equal to 6;

(ii) the intercept on any prism edge by the two asymptotes which meet it is equal to $2\{3+2(\epsilon_1+\epsilon_2+\epsilon_3)\}^{\frac{1}{2}}$;

(iii) the length $OV_1 = 3\{2(1-\epsilon_1)\}^{\frac{1}{2}}$.

The cubical hyperbola will be called *equiangular** when the angles between the asymptotes are all equal.

In this case we have, in the notation of § 54, $\epsilon_1=\epsilon_2=\epsilon_3$. If the asymptotes are mutually perpendicular the cubical hyperbola is called *rectangular*.

In this case ϵ_1, ϵ_2, ϵ_3 are all zero.

59. Rectangular cubical hyperbola.

The rectangular cubical hyperbola has properties analogous to the theorem that the orthocentre of three points on a rectangular hyperbola lies on the rectangular hyperbola.

The direction cosines of the line joining the points with parameters $(\alpha_1, \beta_1, \gamma_1)$, $(\alpha_2, \beta_2, \gamma_2)$ are (§ 43) proportional to

$$\left(\frac{a}{\alpha_1\alpha_2}, \quad \frac{b}{\beta_1\beta_2}, \quad \frac{c}{\gamma_1\gamma_2}\right),$$

the axes of reference being rectangular.

Take six points P_1, P_2, P_3, P_4, P_5, P_6 on the curve and find the condition that the plane $P_1P_2P_3$ is perpendicular to the plane $P_4P_5P_6$.

* The name "equiangular" has been applied by Krüger to a more restricted curve: see § 61.

The direction cosines of a normal to the plane $P_1P_2P_3$ are proportional to the minors of the array

$$\begin{vmatrix} \dfrac{a}{\alpha_1\alpha_2} & \dfrac{b}{\beta_1\beta_2} & \dfrac{c}{\gamma_1\gamma_2} \\ \dfrac{a}{\alpha_1\alpha_3} & \dfrac{b}{\beta_1\beta_3} & \dfrac{c}{\gamma_1\gamma_3} \end{vmatrix}$$

and therefore (§ 32) proportional to $(\alpha_1\alpha_2\alpha_3/a,\ \beta_1\beta_2\beta_3/b,\ \gamma_1\gamma_2\gamma_3/c)$.

The two planes are therefore perpendicular, if

$$\prod_6 \alpha_1/a^2 + \prod_6 \beta_1/b^2 + \prod_6 \gamma_1/c^2 = 0,$$

and the symmetrical form of this relation shows that the plane through any three of the six points is perpendicular to the plane through the remaining three. When five of the points are known the sixth point is in general uniquely determined from the equations

$$\alpha \prod_5 \alpha_1/a^2 + \beta \prod_5 \beta_1/b^2 + \gamma \prod_5 \gamma_1/c^2 = 0, \quad \alpha + \beta + \gamma = 0;$$

but, if

$$\prod_5 \alpha_1/a^2 = \prod_5 \beta_1/b^2 = \prod_5 \gamma_1/c^2,$$

the sixth point is indeterminate, and in this case the line joining any two of the five points is perpendicular to the plane containing the remaining three. Any four of the five points are vertices of a tetrahedron with pairs of opposite edges perpendicular and the orthocentre of the tetrahedron is the fifth point. If the points P_4, P_5 are fixed, $P_1P_2P_3$ is a variable triangle in which the cubic is cut by a system of parallel planes and the locus of the orthocentre of the variable triangle is the chord P_4P_5 (see § 57). The curve therefore has the properties:

(i) given six points on the curve, if the plane through three of these points is perpendicular to the plane through the remaining three, the same is true of *any* two such sets of three points;

(ii) given five points on the curve, if the plane through three of them is perpendicular to the line joining the remaining two, the same is true of any two such sets of two and three points and each of the five points is the orthocentre of the other four.

The rectangular cubical hyperbola presents itself in the geometry of the quadric as the unique twisted cubic through the feet of the six normals from any point to a central quadric; for, taking the quadric as

$$ax^2 + by^2 + cz^2 + d = 0,$$

and the point as P_1, the coordinates of the foot of a normal are given by

$$\frac{x-x_1}{ax}=\frac{y-y_1}{by}=\frac{z-z_1}{cz}=-\frac{1}{\theta}, \text{ say}$$

and therefore $$x=\frac{x_1\theta}{a+\theta}, \quad y=\frac{y_1\theta}{b+\theta}, \quad z=\frac{z_1\theta}{c+\theta}.$$

Taking θ as a parameter, these are the equations of a twisted cubic; it has three real infinity points given by $\theta=-a, -b, -c$ and these infinity points lie on the coordinate axes. The asymptotes of the curve are therefore parallel to the principal axes of the quadric. The value $\theta=0$ gives the centre of the quadric, the value $\theta=\infty$ gives the point P_1: it follows immediately that the six normals are generators of a quadric cone, and that the six lines joining the centre to the feet of the normals are generators of another quadric cone (cf. § 26, 4 (i)).

60. Equilateral equiangular cubical hyperbolas.

The curve is still further specialized if it is both equilateral and equiangular; in this case it has a number of simple metrical properties.

If the point (x_1, y_1, z_1) lies on the curve the cyclic symmetry of the equations giving the circumscribing cylinders shows that the points with coordinates (z_1, x_1, y_1), (y_1, z_1, x_1) also lie on the curve. Now these points lie on the plane $x+y+z-(x_1+y_1+z_1)=0$, parallel to the central plane, and are the vertices of an equilateral triangle, and therefore, if one branch of the curve is given, the other two branches are found by rotating this branch about the central axis through angles $2\pi/3$, $4\pi/3$. Taking three points, one on each branch, on a plane parallel to the central plane, the quadric cones projecting the curve from these points are congruent, and the inscribed conics lying in the osculating planes at these points are also congruent.

The axis of the curve is perpendicular to the central plane; the asymptotes are equally inclined to the central plane; the diameters $\delta_1, \delta_2, \delta_3$ are perpendicular to the chords they bisect; the locus of centres of inscribed conics is a circle; one of the circumscribing revolution quadrics is $yz+zx+xy+9=0$ and has its axis of revolution coincident with the axis of the curve.

Since the prism formed by the asymptotic planes is equilateral, the positions of the asymptotes of any equilateral equiangular cubical hyperbola are determined as follows:

Take any equilateral prism and cut it in a triangle by a plane perpendicular to its edges: through the middle points of the sides of this triangle draw lines in the faces of the prism making equal angles with the plane. These three lines are the asymptotes of the curve: the nature of the curve depends on the angle between the asymptotes and the central plane. The relation of this angle ϕ to the angle η between the asymptotes is given by

$$1 + 2\cos\eta = 3\sin^2\phi.$$

61. Krüger's "equiangular" cubical hyperbola*.

Returning for the moment to the general cubical hyperbola, let us determine under what conditions the conics inscribed in the curve are all rectangular hyperbolas. The equation of the conic in the osculating plane at the point (α, β, γ) is in oblique Cartesian coordinates

$$\frac{U}{\alpha} + \frac{V}{\beta} + \frac{W}{\gamma} = 0,$$

where $\quad U \equiv (-bY + cZ + 1)^2 + 3aX(aX + 4bY + 4cZ)$; etc.

The condition that this is a rectangular hyperbola is linear in the coefficients of the equation, and therefore all the inscribed conics are rectangular hyperbolas if three of them are, say the conics given by the equations $U=0$, $V=0$, $W=0$.

The infinity points of U are given by

$$X=0, \quad aX + 4bY + 4cZ = 0,$$

and these points are conjugate with regard to the isotropic cone

$$x^2 + y^2 + z^2 + 2\epsilon_1 yz + 2\epsilon_2 zx + 2\epsilon_3 xy = 0,$$

if
$$a + 4c\epsilon_2 + 4b\epsilon_3 = 0.$$

Similarly
$$b + 4a\epsilon_3 + 4c\epsilon_1 = 0,$$
$$c + 4b\epsilon_1 + 4a\epsilon_2 = 0,$$

whence
$$\epsilon_1 = -\frac{1}{8}\frac{(b^2 + c^2 - a^2)}{bc}, \text{ etc.}$$

It follows that if $\epsilon_1 = \epsilon_2 = \epsilon_3$, then $a = b = c$ and conversely, so that any curve of this type which is equilateral is also equiangular and conversely.

The curve considered by Krüger may be defined as an equilateral cubical hyperbola having all its inscribed conics rectangular hyperbolas. It has the following properties:

(i) the cosine of the angle between any pair of asymptotes is $-\frac{1}{8}$;

(ii) each asymptote makes an angle $\pi/6$ with the central plane;

* *Zeitschrift für Math. u. Phys.* XXXVIII, p. 344, 1893.

(iii) the osculating planes at the vertices are perpendicular to the asymptotes.

The first two results are immediately verified: for the third we have to find the direction of the normal to the plane

$$-8x + y + z = 0.$$

The pole of this plane with respect to the imaginary circle

$$X^2 + Y^2 + Z^2 - \frac{2\epsilon}{1+\epsilon}(YZ + ZX + XY) = 0$$

has for its equation

$$-(8 + 10\epsilon) X + (1 + 8\epsilon)(Y + Z) = 0,$$

and this is the infinity point of the line $y = 0$, $z = 0$, if $1 + 8\epsilon = 0$.

The curve is determined equivalently by the property that the circular locus of centres of inscribed equiangular hyperboloids (§ 56) coincides with the conic locus of centres of inscribed conics. Comparison of the equations of the two loci leads to the results above.

62. The equilateral rectangular cubical hyperbola.

As a specialized case of the equilateral equiangular hyperbola, we may consider the case where the asymptotes are mutually at right angles. The coordinate system used in §§ 29—53 is then the ordinary rectangular Cartesian system.

As specimens of its metrical properties take the following:

(i) the quadric $yz + zx + xy + 9 = 0$ circumscribing the curve is a revolution form: its imaginary foci are at the ends of the chord lying along the axis of the curve which is also the axis of revolution of the quadric and the axis of the screw containing the tangents.

(ii) the centres of the other revolution quadrics are the points in which the central plane cuts the edges of the triangular prism of asymptotic planes; each axis is parallel to one of the asymptotic planes and makes with the central plane an angle $\sin^{-1}\frac{2}{3}$.

(iii) the curve has three real double normals, the parameters of whose ends are given by the equations

$$2\alpha^2 + (\sqrt{5} - 1)\beta\gamma = 0,$$

$$2\beta^2 + (\sqrt{5} - 1)\gamma\alpha = 0,$$

$$2\gamma^2 + (\sqrt{5} - 1)\alpha\beta = 0.$$

BIBLIOGRAPHICAL NOTE

A full list of authorities is given in

(1) *Royal Society of London Catalogue of Scientific Papers*, 1800—1900. Subject Index. Vol. I. Pure Mathematics.

(2) Loria: *Il Passato ed il Presente delle principali Teorie Geometriche.*

A general discussion on synthetic lines will be found in

(3) Heinrich Schröter: *Theorie der Oberflächen zweiter Ordnung.*

(4) Theodor Reye: *Geometrie der Lage.* Band II.

Some interesting relations between the twisted cubic and the tetrahedral complex are developed in

(5) Sophus Lie: *Geometrie der Berührungstransformationen.* Kap. 8.

Interpretations of the invariants and covariants of a binary cubic in terms of the geometry of a twisted cubic are given in

(6) Grace and Young: *Algebra of Invariants.*

APPENDIX

The verification of the results following is left to the reader

1. *Formulae connected with the twisted cubic given* (§ 22) *by the parametric equations* $x(\theta-a)=y(\theta-b)=z(\theta-c)=t(\theta-d)$.

In §§ 3—17 we found a series of formulae associated with the cubic whose parametric equations are $x:y:z:t=\theta^3:\theta^2:\theta:\frac{1}{3}$, and some of the corresponding formulae were given in §§ 22—25. Further formulae and theorems will now be given.

As in §§ 23—25, we write $f(\theta)\equiv(\theta-a)(\theta-b)(\theta-c)(\theta-d)$.

(i) The equation of the osculating plane at the point with parameter θ is

$$(\theta-a)^3 x/f'(a)+(\theta-b)^3 y/f'(b)+(\theta-c)^3 z/f'(c)+(\theta-d)^3 t/f'(d)=0.$$

(ii) Any quadric circumscribing the curve has equation

$$A\xi+B\eta+C\zeta=0,$$

where A, B, C are arbitrary constants and (see § 25)

$$\xi\equiv(by-cz)(x-t)-(ax-dt)(y-z),\ \text{etc.}$$

One system of generators of this quadric determines on the curve an involution (as in § 10), whose double points are given by

$$A\alpha u+B\beta v+C\gamma w=0,$$

where $\alpha\equiv(b-c)(a-d)$, etc. (§ 22), and u, v, w are the quadratic factors of the sextic covariant of $f(\theta)$ (§ 24).

(iii) The circumscribing quadric cone with its vertex at T has equation $(b-c)yz+(c-a)zx+(a-b)xy=0$.

The equation of the circumscribing quadric cone with its vertex at X is found by writing d, a, t, x for a, d, x, t respectively in this equation.

The equation of the circumscribing quadric cone with its vertex at the point with parameter θ is $A\xi+B\eta+C\zeta=0$,

where $$A=\{f'(d)(\theta-a)^2+f'(a)(\theta-d)^2\}/(a-d)^2,\ \text{etc.}$$

The equation of the inscribed conic in the osculating plane at T is

$$\{(b-d)(c-d)X+(c-d)(a-d)Y+(a-d)(b-d)Z\}^2$$
$$-(\beta\gamma+\gamma\alpha+\alpha\beta)T^2+(b-d)(c-d)(\beta-\gamma)XT$$
$$+(c-d)(a-d)(\gamma-\alpha)YT+(a-d)(b-d)(\alpha-\beta)ZT=0.$$

The equation of the inscribed conic in the osculating plane at X is found by writing d, a, T, X, $-\alpha$, $-\gamma$, $-\beta$ for a, d, X, T, α, β, γ respectively in this equation, and the equations of the inscribed conics in the osculating planes at Y and Z are then found similarly or by cyclic interchange of (a, b, c), (X, Y, Z), (α, β, γ).

(iv) The equation of the quartic developable surface generated by the tangents of the curve is $\xi^2/\alpha+\eta^2/\beta+\zeta^2/\gamma=0$, where ξ, η, ζ are the expressions given in (ii).

(v) The coordinates of any line λ through the point of the curve with parameter θ satisfy the equations (equivalent to one only)

$$A(\theta-a)=B(\theta-b)=C(\theta-c)=D(\theta-d)=E,$$

where

$$E\equiv(bc+ad)lL+(ca+bd)mM+(ab+cd)nN,$$
$$D\equiv(b-c)mn+(c-a)nl+(a-b)lm-alL-bmM-cnN,$$

and $-A$ is derived from D by writing d, a, $-l$, $-N$, M, L, n, $-m$ for a, d, l, m, n, L, M, N respectively, so that

$$A\equiv(b-c)MN+(b-d)lN+(c-d)lM-dlL-cmM-bnN.$$

The expressions for B and C may be derived from D similarly or from A by cyclic interchange of (a, b, c), (l, m, n), (L, M, N).

If λ is a chord of the curve the expressions A, B, C, D, E all vanish; this is equivalent to two independent relations between the coordinates of λ. Any tetrahedral complex which has all the chords of the curve as rays has an equation of the form

$$a'A+b'B+c'C+d'D+e'E=0,$$

where a', b', c', d', e' are arbitrary constants; and the tetrahedral complex of chords associated with the tetrahedron whose vertices have parameters given by the quartic $F(\theta)\equiv\kappa\theta^4+\ldots=0$ has equation

$$\sum_{a,b,c,d} F(a)A/f'(a)+\kappa E=0.$$

Specially the tetrahedral complex $A=0$ has the vertices of its tetrahedron at the points given by $\theta=b, c, d, \infty$, and the tetrahedral complex $E=0$ is associated with the tetrahedron of reference.

(vi) The equations of the two transversals of the four tangents at the vertices of the tetrahedron of reference are, in the notation of § 22,

$$\alpha(\beta-\gamma)\lambda+\beta(\gamma-\alpha)\mu+\gamma(\alpha-\beta)\nu\pm 2\sqrt{3I}(\alpha\lambda+\beta\mu+\gamma\nu)=0,$$

where $I\equiv(\alpha^2+\beta^2+\gamma^2)/24$ is the I-invariant of $f(\theta)$.

(vii) Through each of the four points X, Y, Z, T of the curve can be drawn three lines each meeting two of the tangents at the other three points. Let (XT) denote the line through X which meets the tangents at Y and Z and so on; then the four lines (XT), (YZ), (ZY), (TX) have a single transversal whose equation is $-\lambda+\mu+\nu=0$ (see § 24), and this transversal meets these four lines at the points where they meet the opposite faces x, y, z, t of the tetrahedron of reference. The lines with equations $\lambda-\mu+\nu=0$ and $\lambda+\mu-\nu=0$ have similar properties.

(viii) The parameters of any two points on the curve are given by a quadratic which can be written in the form $pu+qv+rw=0$, where p, q, r are constants and u, v, w are the quadratic factors of the sextic covariant of $f(\theta)$. The equation of the chord joining these points is, in the notation of §§ 22, 23,

$$(q^2+r^2-p^2)\lambda+(r^2+p^2-q^2)\mu+(p^2+q^2-r^2)\nu-2qr\lambda'+2rp\mu'+2pq\nu'=0.$$

In the same way any two chords of the curve are given by equations of the forms $pu+qv+rw=0$, $p'u+q'v+r'w=0$; the screw which (§ 23) contains among its rays these two chords and the four tangents at their ends is given by the equation

$$(qq'+rr'-pp')\lambda+\ldots+\ldots+(qr'+q'r)\lambda'+\ldots+\ldots=0.$$

2. *Miscellaneous theorems on the twisted cubic.*

(i) If in any system of homogeneous coordinates the parametric equations of a twisted cubic are $x:y:z:t=\theta^3:\theta^2:\theta:1$, and the equation of the plane at infinity is $ax+3by+3cz+dt=0$, then the equation of the central plane of the cubic is

$$(a^2d-3abc+2b^3)x+3(abd-2ac^2+b^2c)y+3(2b^2d-acd-bc^2)z$$
$$+(3bcd-ad^2-2c^3)t=0.$$

(ii) There are two points (real or imaginary) from each of which any three lines in space can be projected by three mutually perpendicular planes.

(iii) If P and P' are conjugate points with respect to a twisted cubic (§ 44, (iv)) and P lies on a given line λ, then the locus of P' is another twisted cubic, whose chords are the polar lines of λ with respect to the circumscribing quadrics of the original cubic.

(iv) The locus of the centroid of three corresponding points of three homographic ranges on three lines is a twisted cubic whose asymptotes are parallel to the three lines.

(v) The locus of the centroid of corresponding points of two homographic ranges on a line and on a conic respectively is a twisted cubic, whose asymptotes are parallel to the line and the asymptotes of the conic.

(vi) Chords of a twisted cubic whose ends are corresponding points of two homographic ranges on the curve are rays of a screw. When the two ranges coincide the screw becomes the unique screw containing the tangents.

(vii) If two twisted cubics have two common points and if homographic ranges on the curves have these points self-corresponding, then the joins of corresponding points are rays of a screw.

(viii) The line λ meets the cubical hyperbola of § 28, if

$$9lmn - 3(l+m+n)(mn+nl+lm) + (4l^2 - mn)L + (4m^2 - nl)M$$
$$+ (4n^2 - lm)N + (m+n)MN + (n+l)NL + (l+m)LM - LMN = 0.$$

(ix) The chords of the cubical hyperbola of § 28 are rays of each of the six tetrahedral complexes given by the equations

$$3l(m-n) - (mM - nN) = 0, \quad 3m(n-l) - (nN - lL) = 0,$$
$$3n(l-m) - (lL - mM) = 0,$$
$$3(l-m)(l-n) + l(L+M+N) - MN = 0,$$
$$3(m-n)(m-l) + m(L+M+N) - NL = 0,$$
$$3(n-l)(n-m) + n(L+M+N) - LM = 0.$$

The sum of the first three equations gives an identical zero.

(x) The polar line of the cylinder axis γ_1 (§ 30) with respect to the screw containing the tangents is the harmonic conjugate of γ_1 with respect to the other cylinder axes γ_2, γ_3.

3. *A screw associated with three given lines.*

Three given lines λ_1, λ_2, λ_3 determine a quadric of which they are generators and, if λ_4 is another generator of the same system, any variable generator of the other system cuts these generators in four points P_1, P_2, P_3, P_4 of constant cross-ratio. Writing

$$\alpha = P_2P_3 . P_1P_4, \quad \beta = P_3P_1 . P_2P_4, \quad \gamma = P_1P_2 . P_3P_4,$$

so that $\alpha+\beta+\gamma=0$, the six cross-ratios of the four points are given by the negative ratios of any two of α, β, γ; for instance

$$(P_1P_2P_3P_4)=-\beta/\alpha.$$

(i) For a given cross-ratio determined by (α, β, γ) the equation of the generator λ_4 is $\varpi_{23}\varpi_1/\alpha+\varpi_{31}\varpi_2/\beta+\varpi_{12}\varpi_3/\gamma=0$.

(ii) Two generators give cross-ratios equal to $-\omega$, $-\omega^2$ and these generators are the Hessian lines of λ_1, λ_2, λ_3: they are imaginary when λ_1, λ_2, λ_3 are all real. Their equations are

$$\varpi_{23}\varpi_1+\omega\ \varpi_{31}\varpi_2+\omega^2\varpi_{12}\varpi_3=0,$$

$$\varpi_{23}\varpi_1+\omega^2\varpi_{31}\varpi_2+\omega\ \varpi_{12}\varpi_3=0,$$

and these equations show the Hessian lines as the directrices of the congruence given by $\varpi_{23}\varpi_1=\varpi_{31}\varpi_2=\varpi_{12}\varpi_3$.

(iii) The locus of the centroids of the points in which λ_1, λ_2, λ_3 are cut by a variable transversal is a twisted cubic (§ 51) and the Hessian lines are the tangents to this cubic at the two points lying on its central axis.

(iv) There is a unique screw which has among its rays all the transversals of λ_1, λ_2, λ_3 and also the two Hessian lines of λ_1, λ_2, λ_3. The equation of the screw is $\varpi_{23}\varpi_1+\varpi_{31}\varpi_2+\varpi_{12}\varpi_3=0$. Any two generators (of the same system as λ_1, λ_2, λ_3) which are harmonically separated by the Hessian lines are polar lines with respect to this screw.

(v) Take any three points A_1, A_2, A_3 on λ_1, λ_2, λ_3 respectively: there is a unique twisted cubic passing through A_1, A_2, A_3 and having as its tangents at these points the three transversals of λ_1, λ_2, λ_3. There are ∞^3 such curves, corresponding to different positions of A_1, A_2, A_3, and the tangents of all these twisted cubics are rays of one and the same screw given (iv) by

$$\varpi_{23}\varpi_1+\varpi_{31}\varpi_2+\varpi_{12}\varpi_3=0.$$

Specially, if the points A_1, A_2, A_3 are at infinity, λ_1, λ_2, λ_3 are the axes of the cylinders circumscribing the corresponding cubic.

If λ_4 is any ray of this screw, then $\varpi_{23}\varpi_{14}+\varpi_{31}\varpi_{24}+\varpi_{12}\varpi_{34}=0$, and the relation between the four lines is mutual.

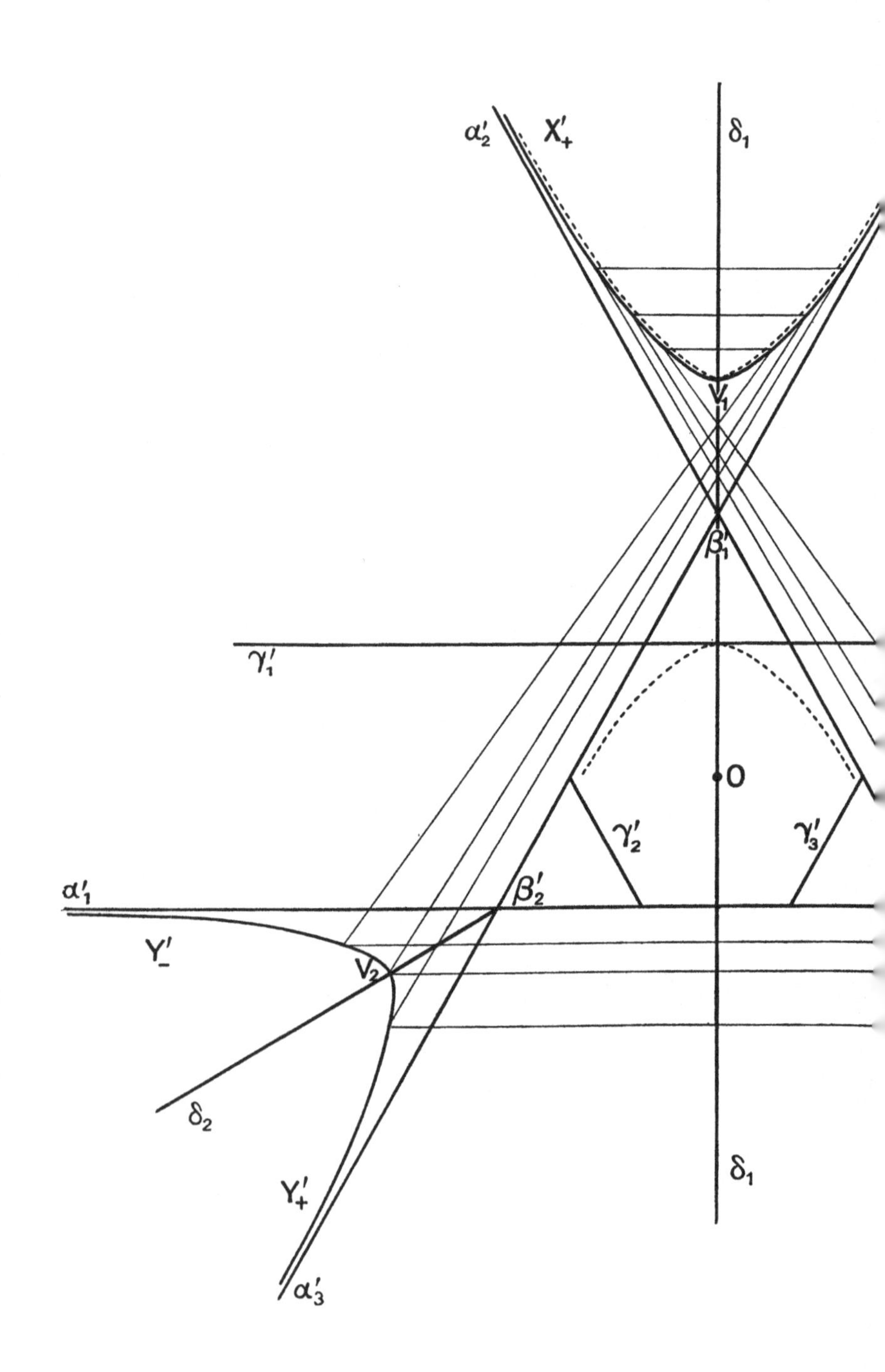

α'₂
X'₊
δ₁
V₁
β'₁
γ'₁
O
γ'₂
γ'₃
α'₁
β'₂
Y'₋
V₂
δ₂
δ₁
Y'₊
α'₃

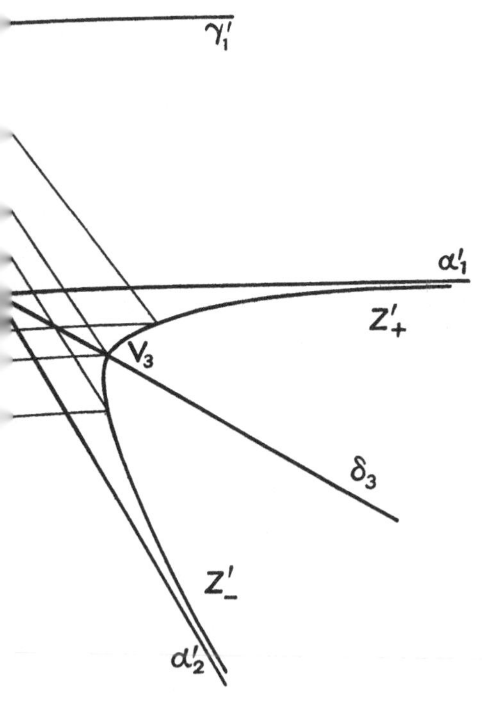

Fig. 10. *Orthogonal projection of an equilateral equiangular cubical hyperbola* (§ 60) *on its central plane, showing two systems of bisected chords* (cf. § 42).

The projection of a_1 is denoted by a_1' and so on; only one-third of the figure is drawn in detail.

(i) The projections of chords bisected by δ_1 are perpendicular to δ_1.

(ii) The projections of chords bisected by γ_1 are tangents to the plane hyperbola shown by the dotted curve; its vertices are V_1 and $\gamma_1\delta_1$, its centre is $\beta_1\delta_1$ and its asymptotes are a_2', a_3'. One branch nearly coincides with $\mathbf{x}'$ and touches it at the vertex: tangents to this branch are projections of chords with real ends. The projections of the chords with imaginary ends touch the other branch with vertex $\gamma_1\delta_1$.

The figure is representative of properties, not involving the imaginary circle at infinity, of the general cubical hyperbola (cf. §§ 28, 54).

Printed in the United States
By Bookmasters